The Practical Guide to Operating and Maintaining a SLA 3D Printer

What this book is all about...

In recent years, the world of 3D printing has experienced a remarkable transformation, transitioning from a niche hobbyist pursuit to a critical tool in various industries. Among the different types of 3D printing technologies, Stereolithography (SLA) stands out for its precision and versatility. Whether you're an artist, engineer, educator, or hobbyist, understanding the nuances of SLA 3D printing can unlock a realm of possibilities, enabling you to create intricate and highly detailed objects that were once impossible to produce without significant time and cost investments.

This book, "A Beginner's Guide to Operating and Maintaining a SLA 3D Printer," aims to demystify the complexities associated with SLA technology. It is crafted to serve as a comprehensive resource for beginners, walking you through every step of the process, from setting up your printer to troubleshooting common issues. Our goal is to empower you with the knowledge and confidence needed to harness the full potential of your SLA 3D printer.

Throughout this book we use the Photon Zero for demonstration.

Book MAJOR REVISION April 2026.

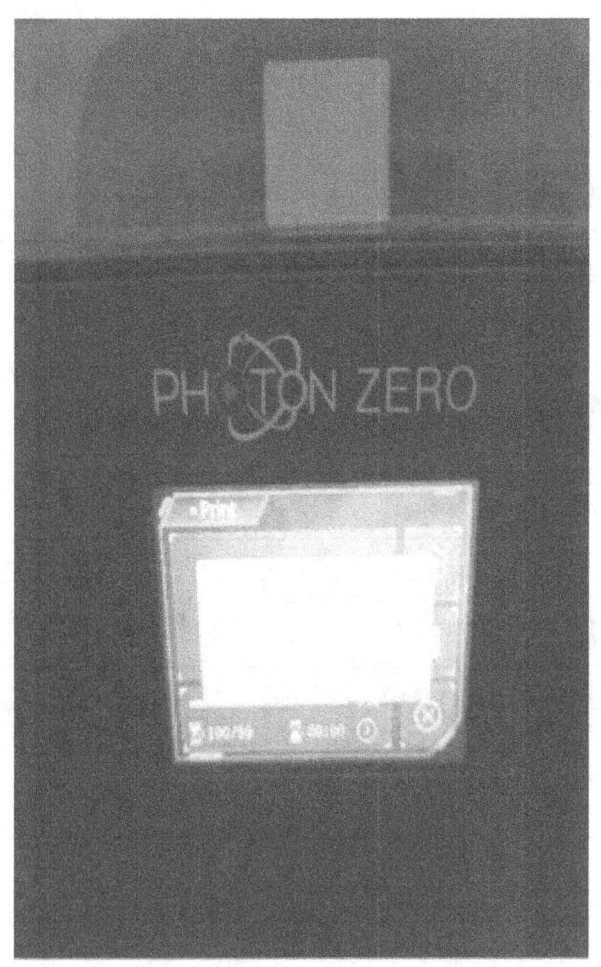

Restrictions on Alteration

You may not modify the Book or create any derivative work of the Book or its accompanying documentation. Derivative works include but are not limited to translations.

Restrictions on Copying

You may not copy any part of the Book unless formal written authorization is obtained from us.

Table of Contents

WHAT THIS BOOK IS ALL ABOUT... 2
INTRODUCTION TO SLA 5
KEY COMPONENTS 10
THE CONCEPT OF A PLANE 15
SLICER 20
RESIN 30
BUILD VS PRINT VOLUME 37
RESOLUTION AND OTHER PARAMETERS 41
PREPARING THE MODEL 48
THE BRAIN OF THE PRINTER 52
INFILL 54
PROPER SUPPORT 55
FILLING, PRINTING AND CLEANING 59
DELAMINATION AND OTHER DEFECTS 69
RETRACTION AND EXPOSURE 74
LEVELLING 77

Introduction to SLA

Stereolithography SLA is one of the pioneering technologies in additive manufacturing, utilizing a process where a UV light source solidifies liquid resin layer by layer to create three-dimensional objects.

To be precise, SLA 3D printing distinguishes itself from other additive manufacturing techniques by using a laser to cure liquid resin into hardened plastic in a layer-by-layer fashion. This method offers unparalleled detail and surface finish, making it the go-to choice for applications requiring high precision, such as jewelry design, dental models, and engineering prototypes.

The process of SLA 3D printing begins with the creation of a digital 3D model using CAD software. This model is then sliced into thin layers using specialized slicing software, converting each layer into a series of commands (G-code) that the printer can interpret. Depending on the desired properties of the final object, an appropriate photopolymer resin is selected.

The actual printing involves a meticulous layer-by-layer construction. Initially, the build platform is positioned just above the bottom of the resin vat, allowing a single layer of resin to sit between the platform and the vat. The UV light source then activates, solidifying the resin according to the G-code instructions for that layer. After each layer is cured, the build platform lifts slightly, making way for a new layer of resin to flow underneath, with the recoating system ensuring an even spread. Each new layer adheres to the previously cured one, gradually building the object.

Once printing is complete, the object is removed from the build platform and undergoes a cleaning process to remove any uncured resin. An additional UV curing process further solidifies the object, enhancing its mechanical properties. The final object may also undergo finishing steps like sanding or painting to achieve the desired surface quality and attributes.

Unlike FDM (Fused Deposition Modeling) printers that extrude filament, SLA printers can produce complex geometries with minimal post-processing, capturing intricate details that are essential for advanced prototyping and artistic endeavors.

The most compelling reason to choose SLA over FDM is when you need high precision and fine detail. SLA can reliably produce features as small as 50 to 100 microns, such as sharp threads, intricate textures, or the delicate features of a miniature figurine, where FDM would struggle or fail entirely. Additionally, SLA prints come out with a smooth, glass-like surface finish right from the printer because the resin cures as a continuous liquid layer. FDM prints, by contrast, show visible layer lines that require significant sanding or chemical smoothing to achieve a similar appearance. For applications like jewelry masters, dental models, or any part where surface aesthetics and tactile smoothness are critical, SLA is the clear winner.

Another major advantage of SLA is part isotropy. FDM prints are inherently weak along the layer lines, meaning they will easily snap or delaminate if force is applied parallel to those layers. SLA prints, however, cure as a nearly homogeneous solid, so their strength is fairly consistent in all directions. This makes SLA preferable for small, functional parts that will endure stress from multiple angles, such as fluidic devices, snap-fit enclosures, or detailed engineering prototypes. Furthermore, SLA offers a much wider range of specialized resin chemistries than FDM does filaments. You can choose from castable resins for lost-wax jewelry casting, biocompatible resins for medical devices, high-temperature resins that withstand over 200 degrees Celsius, or flexible resins that behave like rubber.

That said, SLA has significant drawbacks that explain why FDM remains more common for general use. SLA prints require messy post-processing involving isopropyl alcohol washing and UV post-curing, the liquid resin is toxic and requires gloves and ventilation, and the build volumes are generally much smaller than what even a modest FDM printer offers. The printed parts are also more brittle than typical FDM filaments like PETG or ABS, making SLA unsuitable for load-bearing structural parts, functional hinges, or any object that will experience repeated impact or flexing.

In practical terms, you would reach for SLA when you are making small, detailed, high-quality objects where visual appearance or fine feature resolution is paramount, such as tabletop miniatures, jewelry patterns, dental aligners, prototype lenses, or custom electronic enclosures with snap fits. You would choose FDM when you need large parts, tough and durable mechanical components, something inexpensive to prototype quickly, or any object where being able to sand or drill the part after printing is acceptable. Many professionals keep both types of printers, using FDM for large structural components and SLA for the fine details that FDM cannot reproduce.

This guide is tailored for those new to the world of SLA 3D printing, yet it is also a valuable resource for those looking to deepen their understanding of this sophisticated technology. Whether you're setting up your first SLA printer or seeking to refine your maintenance skills, you will find practical tips, detailed explanations, and step-by-step instructions to help you navigate the learning curve.

Key components

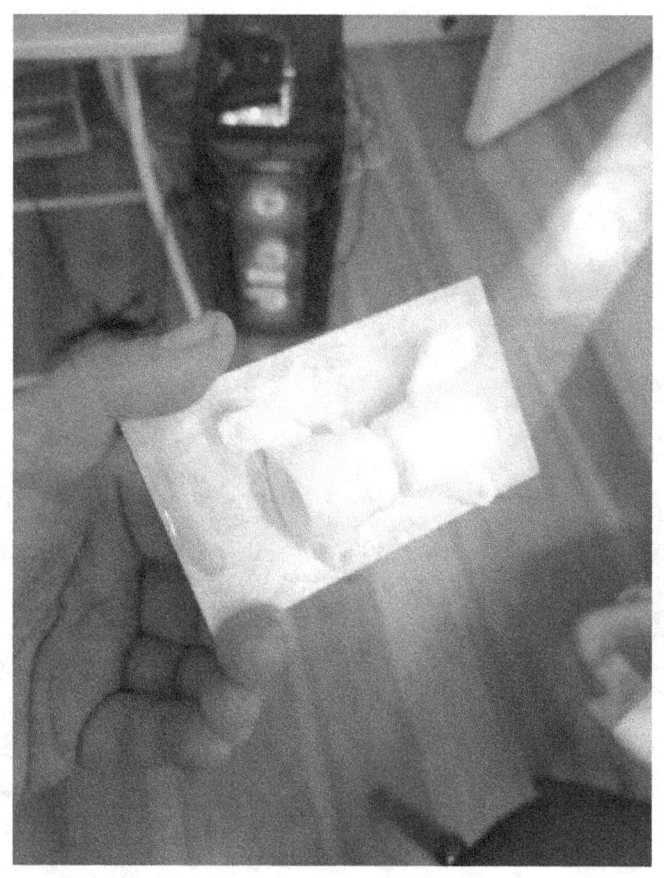

The resin tank, or vat, is a transparent container that holds the liquid photopolymer resin, which is crucial for the 3D printing process. Typically constructed from materials like acrylic or glass, the tank allows UV light to penetrate it, facilitating the curing process. Its bottom is often coated with a non-stick layer such as PDMS or FEP, which enables the easy separation of cured layers from the tank. The tank acts not only as a reservoir for the resin but also as the initial platform where the curing begins.

The build platform serves as the surface on which the 3D object is constructed. Made from robust materials such as aluminum, the platform moves vertically, facilitating the layer-by-layer building of the object. Initially, it submerges into the resin, starting just above the tank's bottom, and with each layer's completion, it incrementally rises to allow for the addition of new resin layers.

Central to the SLA process is the UV light source, which solidifies the resin by exposing it to ultraviolet light. Various types of UV light sources are employed, including solid-state lasers (often with a 405 nm wavelength) for precision, DLP projectors that project entire layers at once for efficiency, and LED arrays for uniform exposure. The primary function of the UV light source is to selectively cure the resin at specified points, dictated by the 3D model being printed.

To achieve precise movements during printing, SLA printers use a combination of a gantry system and a galvanometer (galvo) system. The gantry system is a mechanical framework that moves the UV light source or build platform along the X, Y, and Z axes. The galvo system, on the other hand, consists of rotating mirrors that swiftly direct the laser beam across the resin surface, enabling accurate curing through rapid adjustments in the beam's direction.

To ensure each new layer of resin is evenly spread over the previously cured layer, some advanced SLA printers incorporate a resin recoating system. This system, which may use a mechanical blade or a wiper, spreads a fresh, uniform layer of resin, ensuring consistent thickness and layer formation throughout the printing process.

The control system, typically comprising a microcontroller or a computer, serves as the printer's brain. It processes G-code or other command sets to direct the UV light source, build platform, and other components. Additionally, various sensors monitor and regulate critical parameters like temperature and resin levels, ensuring the printing environment remains optimal.

SLA printers are significantly sensitive to their surrounding environment. Temperature is one of the most critical factors, with an ideal printing range of roughly 20°C to 30°C (68°F to 86°F). If the temperature drops too low, the resin becomes more viscous and flows poorly, which can lead to under-exposure, delamination, or complete print failure. Conversely, if the temperature is too high, the resin may cure too quickly, creating internal stresses that cause warping or deformation. Even after printing, cured parts can soften or deform when exposed to temperatures around 85°C, though high-temperature resins are available for more demanding applications. To maintain a stable temperature, it is wise to use a heated enclosure during cold months and to keep the printer away from windows, AC vents, or direct heaters.

Humidity also plays a significant role, with an ideal relative humidity range of 20% to 50%, and many users recommend keeping it strictly below 40% to 50%. When humidity is too high, uncured resin absorbs moisture from the air, resulting in soft, sticky prints with weak layer adhesion and surface bubbles. For fully cured parts, prolonged exposure to high humidity degrades mechanical properties, making the parts more brittle and reducing their overall strength. In humid climates or basements, using a dehumidifier is a worthwhile investment, and resin bottles should always be stored tightly sealed.

Air currents and drafts from fans or air conditioning can cause uneven cooling and temperature gradients around the printer, which in turn affects resin viscosity, FEP tension, and overall machine precision. These unpredictable changes often lead to print failures. While drafts have little direct impact on already cured parts, it is best not to aim fans or AC vents directly at the printer. If a drafty vent cannot be avoided, even a simple cardboard shield can help block the airflow.

Ambient UV light is another concern, since uncured resin left in the vat will harden when exposed to any UV source, including sunlight through a window. For finished parts, prolonged outdoor UV exposure will embrittle them over time, turning tough prints into brittle ones. Placing the printer away from windows is essential, and for parts intended to live outdoors, applying a UV-blocking clear coat, such as automotive clear coat, can significantly extend their lifespan.

To maintain operational temperatures and prevent overheating, especially of the UV light source, some larger scale SLA printers utilize cooling systems. These can be either air or liquid cooling systems, and they play a crucial role in ensuring consistent print quality and prolonging the printer's lifespan.

Encasing the printer serves multiple purposes: it shields users from UV exposure, protects the printing area from contaminants like dust, and often includes UV-filtering capabilities to safeguard against potential light leakage.

SLA 3D printing boasts several advantages, including high resolution and accuracy, making it ideal for producing fine details and smooth surfaces. The technology offers material versatility, with a range of resins available that can provide different mechanical properties and aesthetic options. Additionally, SLA printers can produce complex parts relatively quickly, compared to traditional manufacturing methods. However, there are limitations to consider. The process is confined to using photopolymer resins, which can be brittle and lack heat resistance compared to other materials. Post-processing is often required to clean and cure the parts, which can add time and complexity to the workflow. Additionally, SLA printers typically have size limitations, with smaller build volumes compared to other additive manufacturing technologies.

The concept of a plane

Understanding the concept of the X, Y, and Z axes in SLA 3D printing is fundamental to appreciating how these machines create intricate and detailed objects layer by layer. The precise control and coordination of movements along these axes by the printer controller are critical for the quality and accuracy of the printed objects, making them an essential aspect of the 3D printing process.

From a design perspective, a plane (aka work plane) is like a table top — it gives you a surface for working.

Things "on" a plane can be moved or scaled in multiple directions. Movement on the plane is movement on the Z surface, which is front to back. On the other hand, X is side to side while Y is up and down.

The X and Y axes in an SLA 3D printer correspond to the horizontal plane, controlling the lateral movements of the laser or light source as it solidifies the resin.

There is no mechanical XY movement since there is no such thing as "print head" in a SLA printer (FDM printer does have print head)..

The X axis runs horizontally from left to right when viewed from the front of the printer. It dictates the horizontal positioning of the laser or light source across the width of the resin vat. As the laser moves along this axis, it defines the width of the layer being cured. In contrast, the Y axis runs horizontally from front to back. It determines the horizontal positioning of the laser or light source across the depth of the resin vat. Movement along the Y axis defines the depth of the layer being cured. In combination, the X and Y axes allow the laser to trace complex shapes and patterns on the surface of the resin. This coordinated movement defines the two-dimensional outline of each layer of the 3D object. The precision and accuracy of movements along these axes are crucial for achieving detailed and accurate prints.

The X and Y axes directly impact the resolution and detail of the printed object. Higher precision in these axes allows for finer details and smoother surfaces, as the laser can more accurately follow the intricate patterns defined by the 3D model. In SLA printers, resolution in the X and Y directions is often defined by the spot size of the laser or the pixel size of the light source, with smaller sizes enabling higher detail.

The Z axis is responsible for the vertical movement, controlling the elevation and lowering of the build platform. Simply put, the Z axis runs vertically, from the bottom of the printer towards the top. It moves the build platform up and down, allowing the printer to build the object layer by layer. After each layer is cured, the build platform rises slightly (or the resin tank lowers in some designs), making room for the next layer of resin to be spread over the previously cured layer. In other words, each increment along this axis corresponds to the thickness of a single layer of the object. The precision in Z axis movement ensures that each new layer is positioned correctly relative to the previous one, maintaining the overall structural integrity and accuracy of the final object.

The Z axis affects the layer thickness, which plays a significant role in the surface finish and vertical resolution of the printed object. Thinner layers result in smoother surfaces and more accurate vertical details, as each layer contributes less to the overall height and is less noticeable in the final product. However, thinner layers also mean longer print times since more layers are needed to complete the object. Accurate movements along the Z axis are vital for ensuring that each layer adheres properly to the one below it. Any errors in Z axis positioning can lead to misalignment between layers, resulting in weak spots or deformations in the printed object. This is especially crucial for tall or complex structures where even minor misalignments can accumulate and significantly affect the final result.

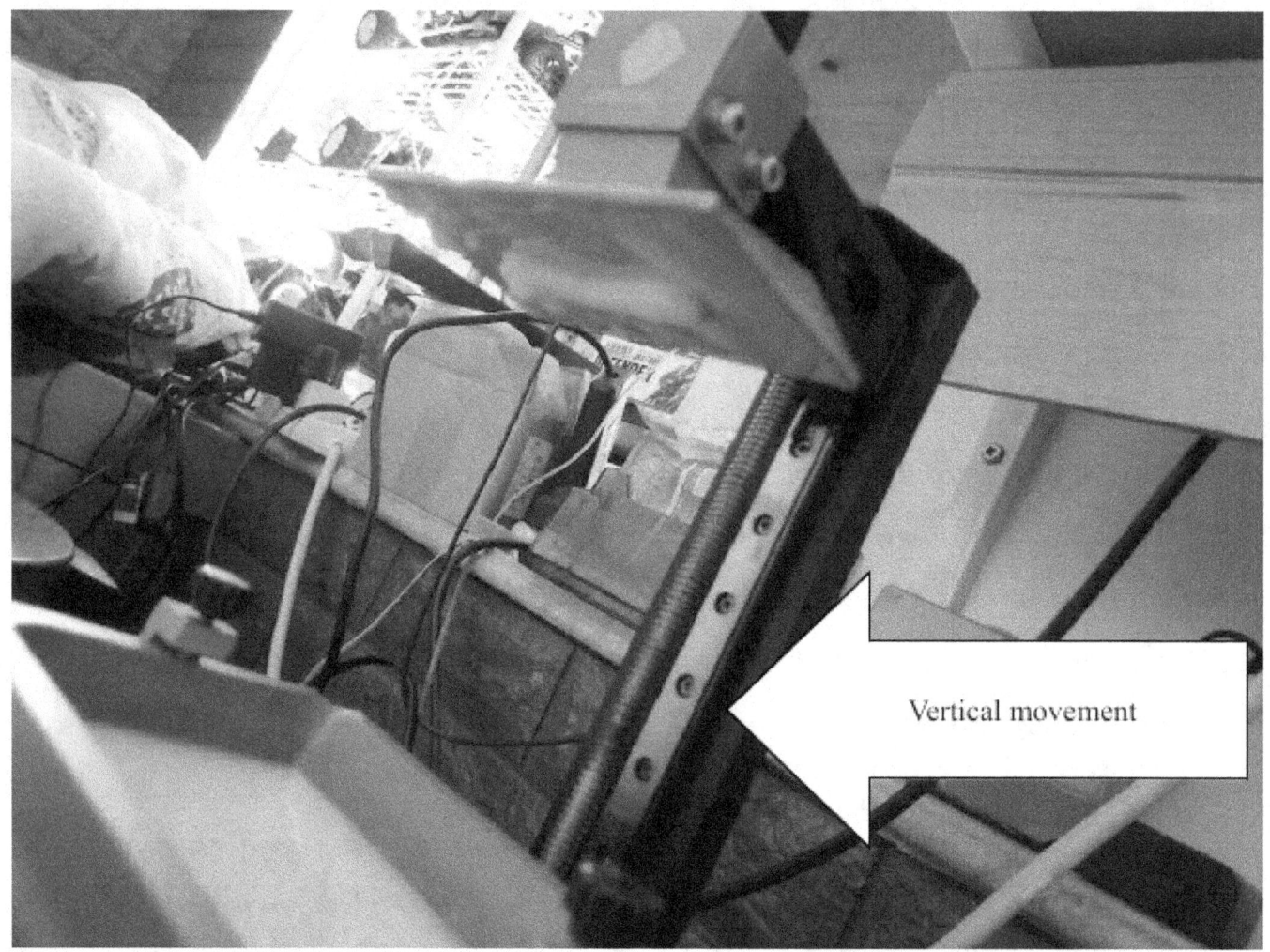

Vertical movement

Regular calibration of the X, Y, and Z movements is essential to maintain print accuracy. Any deviations or mechanical issues in these axes can lead to errors in the printed object, such as incorrect dimensions, layer shifting, or surface defects. Ensuring that the printer's movements are well-calibrated and smooth helps achieve consistent and high-quality prints.

The LED control panel of the printer would allow you to manually control movement (Z axis only) for adjustment or other maintenance purpose.

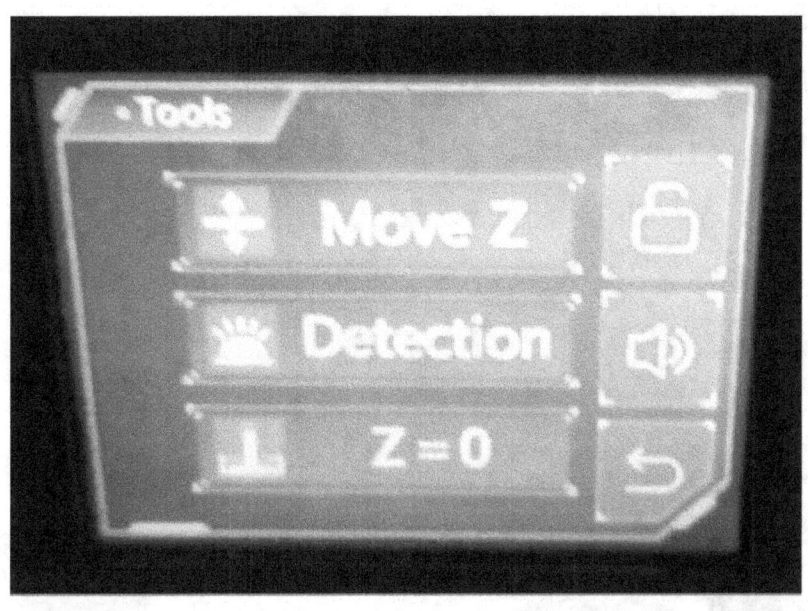

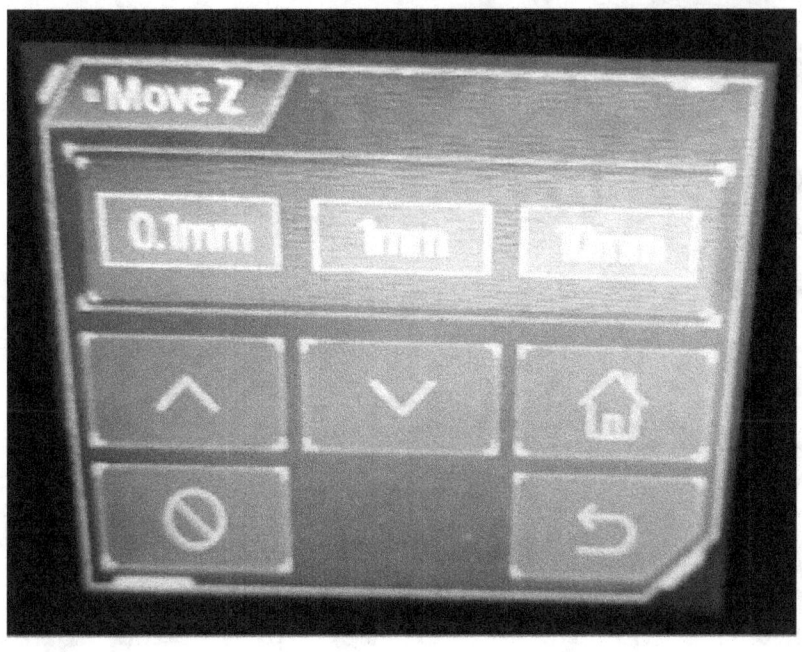

Slicer

The entire 3D printing process is governed by digital instructions typically encoded in G-code, which specifies the exact movements along the relevant axes. Slicing software converts the 3D model into a series of these instructions, dividing the model into thin layers and defining the path the laser should follow for each layer. The software ensures that the movements along each axis are precisely coordinated to replicate the 3D model accurately.

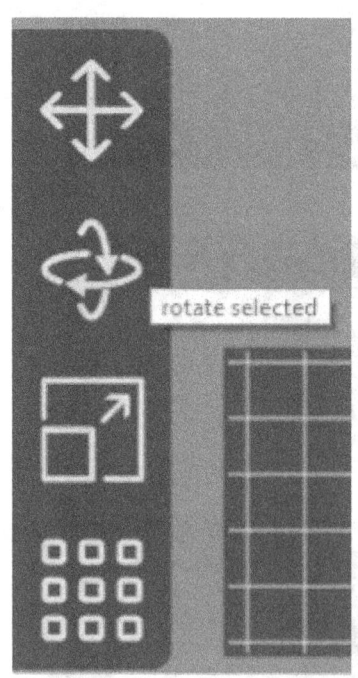

Regardless of the type of printer you use, the slicer still works the same. Your slicer software would allow you to move and rotate your object along X, Y or Z, allowing you to determine the print position relative to the print bed. Some even allow you to scale or mirror your object right before printing. How the 3D printer implements the model and performs the printing is not something you need to worry about.

A slicer is a software that prepares a 3d model for printing. The process is calling slicing. Keep in mind, the one that comes with your printer may not be the latest version. If you upgrade to a new version, you must make sure the new version has all your printer specific settings.

In the case of our demo printer, a custom slicer is provided. It is called Photon Workshop.

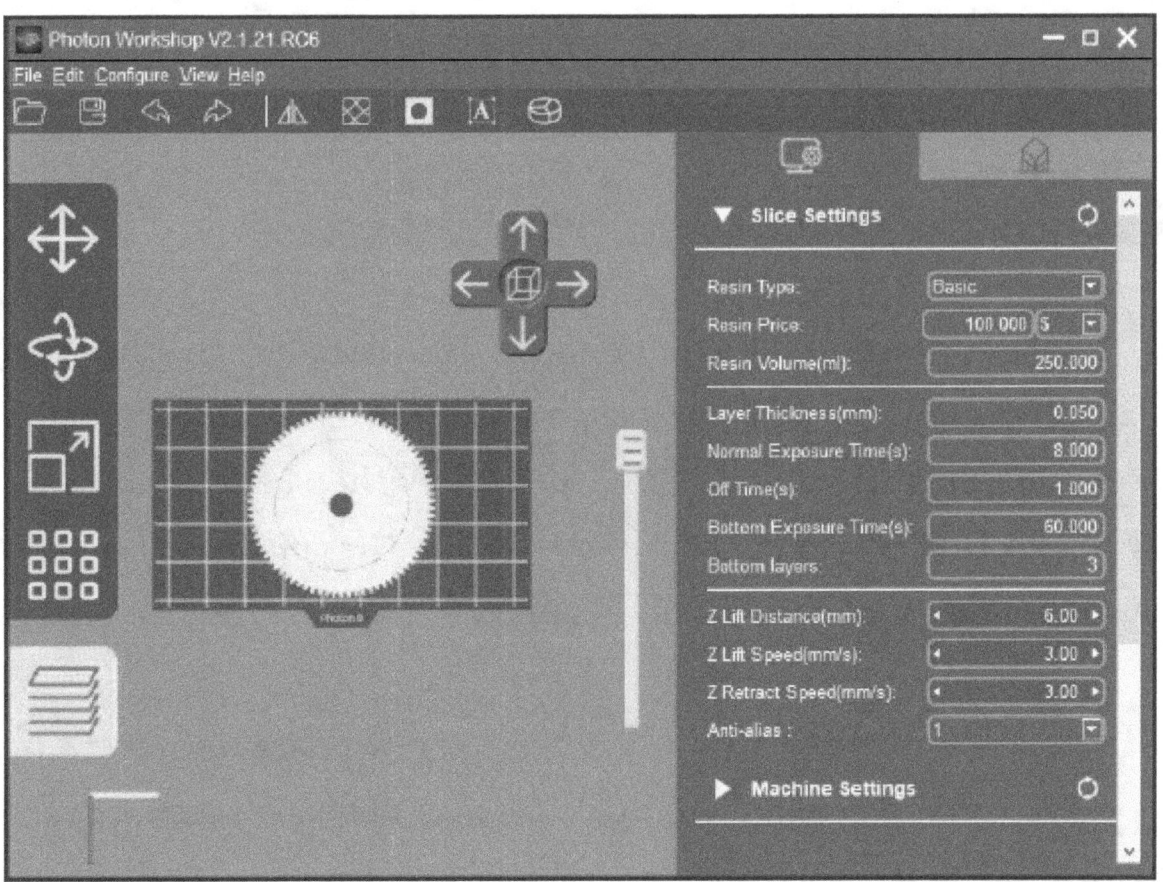

Machine settings and print quality expectations together would form a print profile for keeping 3d printer related settings on print quality. It is printer specific (that's why the slicer software that comes with your printer would have this profile already included). If you are using a non-supported platform, you may need to configure the profile manually (the machine setting must be correct). In any case, you must choose the correct printer model and set the resin type based on manufacturer's instructions.

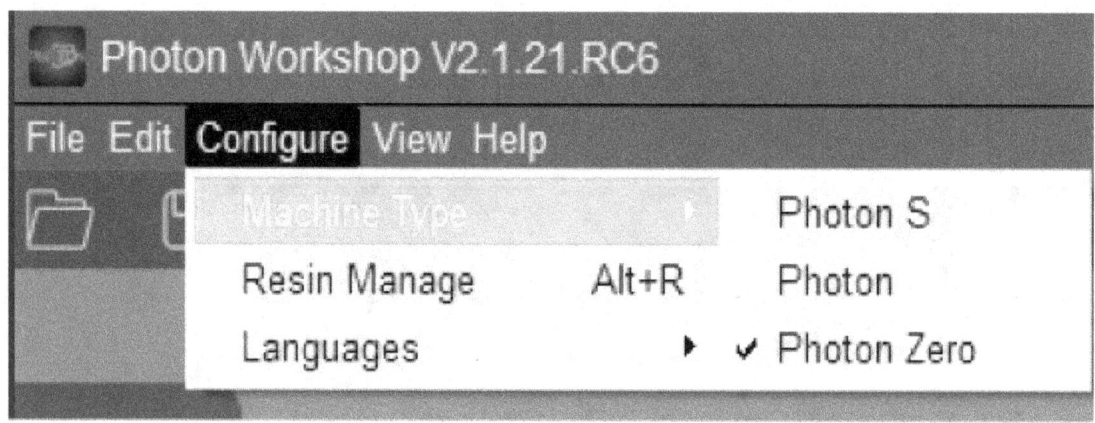

In general, a proper profile should include information on the bed size. The bed shape and size is directly related to the size of object your printer can print. On a SLA printer, the bed is always rectangular (unlike FDM, it is never circular). The parameters are machine level settings that can be fine tuned:

If the printer profile in your slicer is set to a print size larger than your SLA printer's actual build volume, several problems can occur, ranging from software errors to physical damage. The most immediate and common issue is that the slicer will generate toolpaths and slice files for areas that do not exist on your printer. When you transfer the file to the printer and attempt to print, the printer's firmware will typically detect that the requested coordinates exceed its physical limits. This usually results in an error message on the printer's screen, and the print will abort before any resin is exposed. In some cases, the printer may attempt to move its build plate or laser galvanometer beyond their intended ranges, causing the motors to skip steps or make a grinding noise as they hit their mechanical stops.

Beyond simple error messages, setting the wrong build volume can lead to partial or shifted prints. The printer might begin the job correctly, but when it reaches a layer that extends beyond the actual screen or laser reachable area, the motor controlling the build plate or the mirror galvanometer may stall, causing the entire subsequent print to be misaligned. This results in a print that is physically warped, has a distinct shear line where the shift occurred, or contains missing sections. The print will often be pulled off the build plate at that point due to the abnormal forces.

There is also a risk of physical damage to the printer itself, particularly for laser-based SLA machines. If the slicer commands the laser to fire at a position outside the resin vat's intended area, the laser beam could potentially strike the interior housing of the printer or the edges of the vat. This can burn or scar non-replaceable components, damage optical mirrors, or melt the plastic frame of the vat. For MSLA printers that use an LCD screen, commanding a larger build area means the screen mask will attempt to display pixels that do not exist, which generally just results in an error, but prolonged attempts could theoretically stress the controller board.

Finally, a less obvious but frustrating issue is that the slicer may proceed and generate a slice file without error if the model is small enough to fit entirely within the incorrectly defined larger volume. In this scenario, the print will be centered based on the wrong coordinates, meaning your model could end up printing far off-center on the build plate, possibly hanging partially over the edge of the actual screen or vat area. This leads to a successful slice file that consistently produces failed, offset prints without any clear error message, making the problem difficult to diagnose.

Machine Settings	
Resolution(pixels):	480x854
XY-Pixel size(um):	115.500
X size(mm):	54.000
Y size(mm):	96.997
Z size(mm):	150.000

When the object is not properly "placed" on the bed, the slicer will have nothing to slice. You can move the object manually to make sure it sits on the bed.

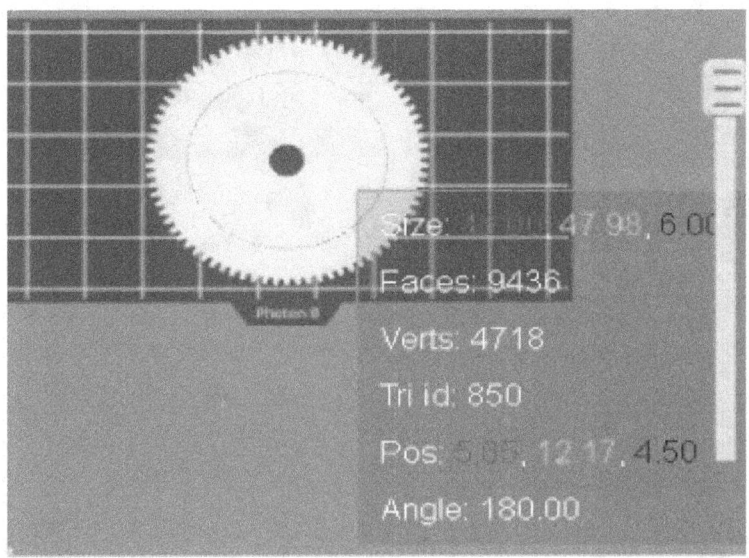

The actual placement of the object can be confusing. See this example:

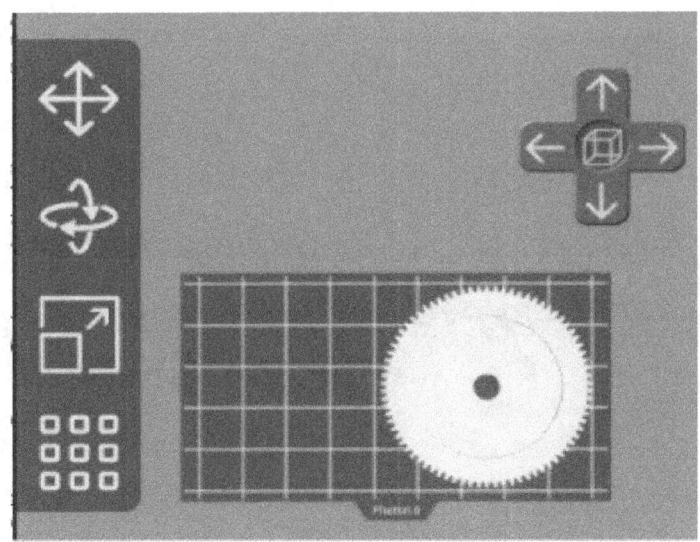

On the software it appears the object is placed on the right. However, in reality it is on the left facing the user.

FYI, there are other slicers on the internet. In our example, however, only the manufacturer supplied slicer is supported due to the need for the proprietary .pw0 file format. STL file must first be converted to pw0 in order to be printed.

While CHITUBOX is the most popular, Lychee Slicer is its main competitor and is also highly regarded, especially among users who prioritize a modern, workflow-optimized experience. Lychee Slicer is often described as more intuitive and easier to learn than CHITUBOX, thanks to features like its "magic" button which can automatically orient and support a model for printing. It is particularly favored by those printing detailed objects such as miniatures or jewelry, as well as by users who need to batch process many models at once for small-scale production. The choice between the two often comes down to a trade-off: CHITUBOX is generally considered the best all-around, free option with the broadest printer support, while Lychee Slicer is known for having a more polished and user-friendly interface, though some of its more advanced features are locked behind a paid subscription.

Chitubox Basic is also free.

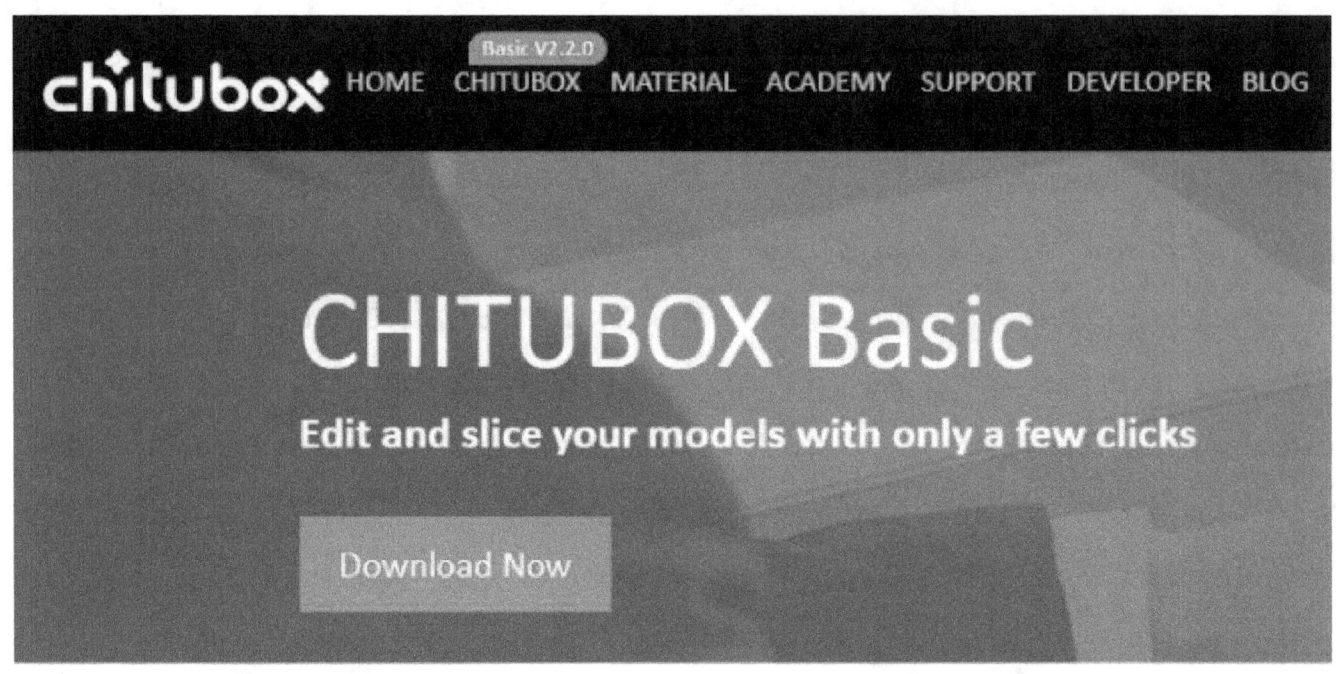

There are also other notable slicers, though they are often more specialized. For instance, PrusaSlicer is a free, open-source option that supports resin printing, and Anycubic Photon Workshop is a dedicated, free slicer for Anycubic printer owners. For most people starting with resin printing, however, CHITUBOX and Lychee Slicer remain the two primary options to consider.

Cura can work with SLA printers, but this capability has important limitations that users should understand. While Cura is best known as the dominant slicer for FDM filament printers, it does offer an experimental resin slicing mode that was introduced in version 5.0 around 2022. This mode provides basic functionality for SLA and DLP printing, including exposure settings for bottom and normal layers, support generation, and layer preview features. Additionally, several manufacturers have historically used Cura as the basis for their SLA printer software, with examples including the StoneFlower

Riverside printer from around 2015 and the Sharebot Antares, and a GitHub repository also exists for Moai SLA printer profiles designed for Cura, though that dates back to 2017.

However, for most users, Cura is not the recommended choice for resin printing. Its resin mode remains experimental, meaning it lacks several critical features that dedicated resin slicers offer.

Resin

Apart from the printer itself, make sure you setup the resin information properly. Follow the resin specific information based on supplier recommendations. OR, use custom type if there are special considerations on the resin you use.

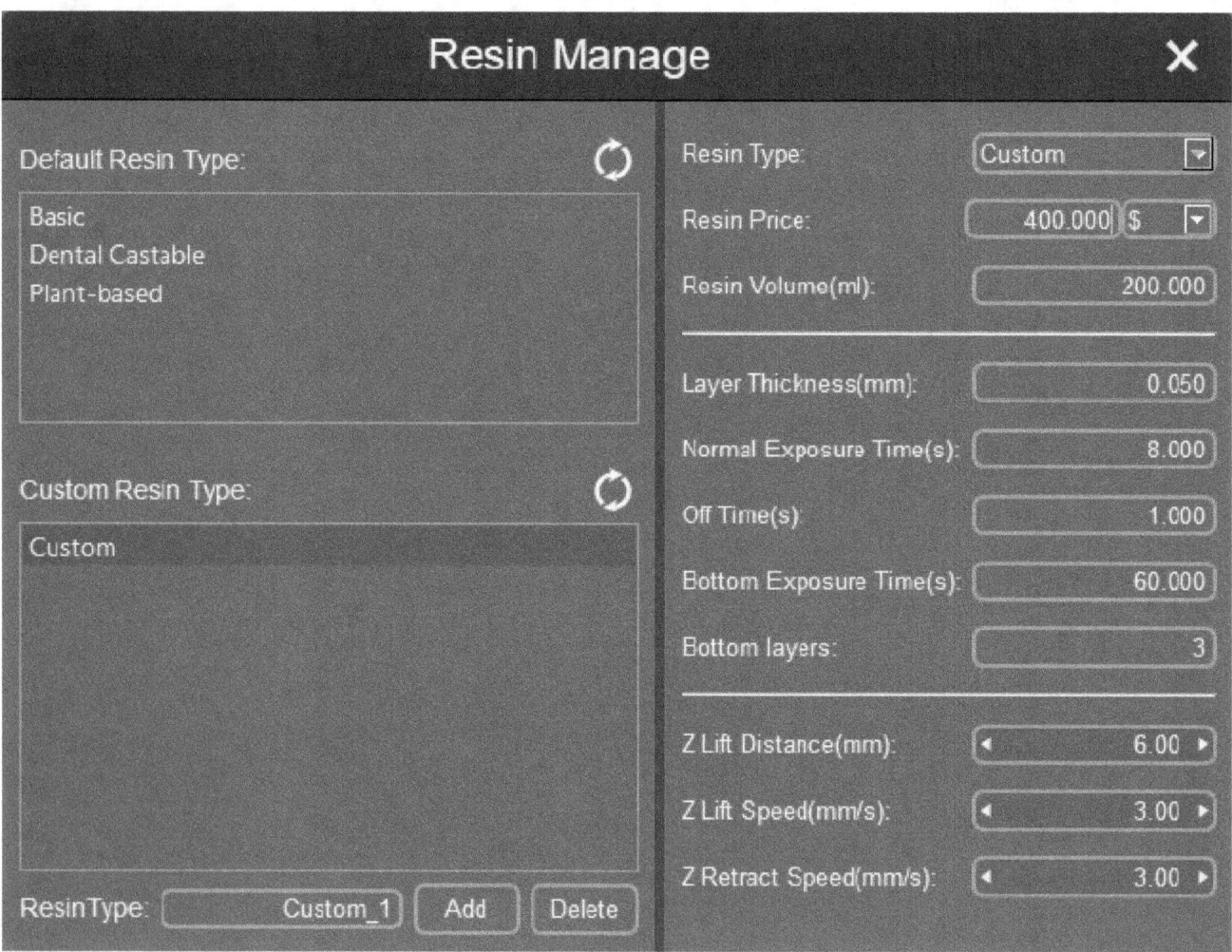

If you do not have Cura, there are other free alternatives. For example: Slic3r is open source and free. Craftware is another.

Resin Price:	100.000 $
Resin Volume(ml):	250.000
Layer Thickness(mm):	0.050
Normal Exposure Time(s):	8.000
Off Time(s):	1.000
Bottom Exposure Time(s):	60.000
Bottom layers:	3
Z Lift Distance(mm):	6.00
Z Lift Speed(mm/s):	3.00
Z Retract Speed(mm/s):	3.00
Anti-alias:	1

Standard resins are the most frequently used in SLA printing due to their balanced combination of cost, ease of use, and performance. They are ideal for general-purpose applications, producing parts with a smooth surface finish and high detail resolution. However, standard resins are relatively brittle, which makes them less suitable for functional parts that require significant impact resistance or flexibility. These resins come in a wide range of colors and can often be post-processed to enhance their aesthetic qualities, making them ideal for prototyping, visual models, and concept validation where mechanical strength is not a critical requirement.

Tough resins are designed to produce parts with higher impact resistance and improved mechanical properties compared to standard resins. They are less brittle and can withstand more stress and strain, making them suitable for applications that demand durability. Tough resins often mimic the properties of ABS plastic and are available in muted colors like gray, black, or clear. These resins are commonly used for functional prototypes, mechanical parts, and other applications where the printed parts need to endure mechanical stress and wear.

Flexible resins create parts with rubber-like qualities, allowing them to bend, compress, and return to their original shape. These resins are typically softer and can range from very flexible to semi-rigid, with properties that include lower hardness and higher elongation. Flexible resins are usually available in opaque or translucent finishes and are ideal for applications that require elasticity, such as wearable items, seals, gaskets, and other parts requiring flexibility and resilience.

These are some examples of popular resins on the market:

- Formlabs Grey Resin is a versatile resin which is ideal for general-purpose prototyping and parts requiring fine details. It is often used for creating models and prototypes where precision and surface finish are important.

- Formlabs White Resin is perfect for creating parts with a clean, white finish, suitable for applications like concept models and parts that need to be painted or finished with a uniform color.
- Formlabs Clear Resin is often used for printing parts that require optical clarity or need to simulate glass or other transparent materials.
- Formlabs Black Resin offers a smooth, matte finish and is ideal for showcasing details and textures, often used in applications such as miniatures or parts that need a sleek, black appearance.
- Anycubic Grey Resin is a popular choice for prototyping, offering a balance of durability and detail, with a neutral grey color that highlights fine details.
- Anycubic Transparent Green Resin provides good transparency and is often used for creating parts that need to demonstrate internal structures or for decorative purposes.
- Anycubic White Resin is often used for printing parts that need a clean, crisp finish, and is easy to paint or dye.
- Elegoo Grey Resin offers excellent detail and surface finish, making it ideal for creating highly detailed prototypes and models.
- Elegoo Transparent Resin is often used for printing parts that require some level of translucency, such as lenses or light covers, while providing a smooth surface finish.
- Elegoo Black Resin is perfect for parts needing a dark, consistent finish, often used in miniature models and functional prototypes.

- Monocure Rapid Model Resin is designed for rapid prototyping as it cures quickly and offers a smooth finish, making it ideal for detailed models and quick-turnaround projects.
- Monocure Clear Resin provides high transparency and is suitable for applications where clarity and a glass-like finish are required.
- Phrozen Aqua Grey 4K Resin is specially formulated for high-resolution prints, this resin offers excellent detail and a smooth, matte finish, making it ideal for creating intricate models and prototypes.
- Phrozen Aqua Clear Resin provides a good level of transparency and is often used for parts that need to demonstrate internal details or for aesthetic applications requiring a clear finish.
- Siraya Tech Fast Grey Resin offers a quick curing time and excellent detail resolution, ideal for general-purpose prototyping and models requiring a fine surface finish.
- Siraya Tech Fast Clear Resin provides clarity and is often used for parts that need a translucent appearance or for applications where internal visibility is important.
- Nova3D Standard Grey Resin is ideal for creating detailed prototypes and models and can provide a neutral grey color that enhances the visibility of fine features.
- Nova3D Standard Transparent Resin offers good clarity and is suitable for applications that require a clear or translucent finish, such as optical parts or decorative items.

SLA resin is a photosensitive material that can degrade or polymerize when exposed to light or improper conditions, which can affect its performance in printing. To sum up:

- SLA resin is highly sensitive to UV and visible light. Exposure can initiate the curing process, causing the resin to harden and become unusable. Store resin in opaque or dark-colored bottles that block out light. Original resin containers are usually designed to be light-proof. Keep the resin in a cabinet or a dark storage room away from any light sources, including sunlight and artificial lighting. Close the resin container immediately after use to minimize light exposure.
- Temperature fluctuations can alter the viscosity of the resin and its chemical stability. Extreme temperatures can lead to resin degradation or make it unusable. Store resin at a consistent, cool room temperature, typically between 15°C and 25°C (59°F to 77°F). Avoid temperatures below 10°C (50°F) and above 30°C (86°F). Never allow resin to freeze, as this can alter its chemical properties and render it unusable. Keep resin away from heat sources like radiators, stoves, or direct sunlight that could cause temperature spikes.
- High humidity can lead to moisture absorption by the resin, which can affect its performance and cause print failures. Store resin in a dry area with low humidity, ideally below 50%. Place silica gel packets or other desiccants in the storage area to absorb any

excess moisture. Always ensure that resin containers are tightly sealed to prevent moisture ingress.

Can you leave unused resin in the tray (aka the resin vat or resin tank)? The short answer is yes, if not for too long (several days would be ok) and if not exposed to day light. Partially used resin in the vat can become contaminated or degrade if not stored properly. To leave resin in the vat for extended periods, cover the vat with an opaque lid to prevent light exposure and contamination. If you want to pour the leftover back to the bottle, filter the resin to remove any cured particles before pouring it back.

Build VS Print Volume

Build volume = W × D × H. Our demo printer is LCD-based. The build volume is 97 x 54mm x 150mm but the printing volume is 97 x 54mm x 150mm.

The build volume defines the maximum size of the object it can produce and is determined by a combination of several factors. Firstly, the resin vat size plays a crucial role, as it is the container that holds the liquid photopolymer resin. The dimensions of the vat (the tray that holds the resin), specifically its width and depth, set the horizontal limits of the build volume, while the height of the vat, coupled with the vertical travel capacity of the build platform, determines the maximum vertical height of the object that can be printed. In our example the container is removable.

The build platform size is another critical component, as it provides the surface on which the object is built. The platform's surface area, dictated by its length and width, must accommodate the base dimensions of the object.

 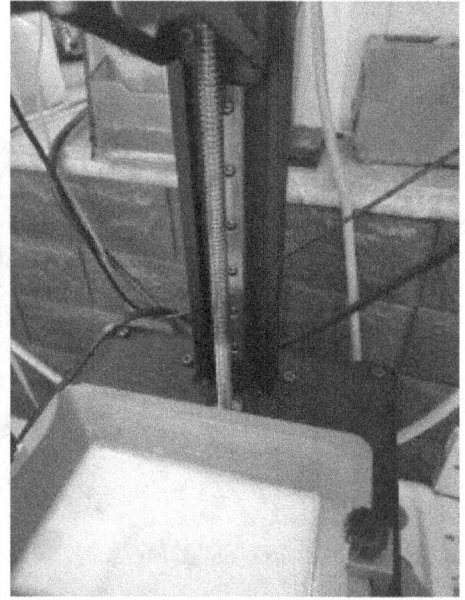

The vertical capacity of the build platform, which refers to its range of movement along the z-axis, further defines the maximum possible height of the object. The z-axis mechanism, which controls the vertical movement of the build platform, significantly influences the build volume. The maximum travel distance of this mechanism sets the height limit for the printed object. The precision and stability of the z-axis mechanism are essential for maintaining alignment, especially for taller prints.

The optical system of the SLA printer, which may use a laser or a projector, determines the method and area of resin curing. The size of the laser spot or the resolution of the projector affects the detail and size of the printed area. For laser-based systems, the galvanometers that direct the laser must cover the entire build area.

In the case of projector-based systems, the light source needs to uniformly cover the resin vat to ensure consistent curing across the build surface.

Mechanical and structural constraints of the SLA printer also impact the build volume. The printer's frame must be large and sturdy enough to accommodate the resin vat and build platform while allowing sufficient movement. Larger builds necessitate a robust frame to prevent vibrations and maintain the accuracy of the print.

The software and firmware of the SLA printer are also determining factors. They must be capable of defining and managing the build area, and the slicing software must efficiently handle data for larger prints. This ensures that the printer can accurately process and produce objects within the desired build volume.

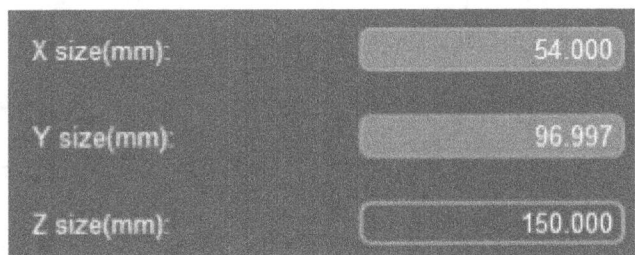

Heat management is another vital consideration. SLA printers generate heat during operation, particularly when producing larger prints. Effective cooling systems are essential to prevent warping or curing issues, thereby indirectly limiting the build volume by ensuring stable printing conditions.

Lastly, the properties of the resin used in the printing process can affect the maximum build volume. Different resins have varying curing characteristics, such as curing depth and speed, which can influence the maximum height or overall volume of the printed object.

Definition-wise, printing volume refers to the actual space or volume occupied by the printed object or objects within a single print job. It is a dynamic measurement that can vary depending on the specific print job and arrangement of the objects being printed. Simply put, it is concerned with the current print job and represents how much of the build volume is actually being used. It is a practical measure of the material usage and efficiency of a particular print job. Therefore, unlike build volume, the printing volume can change from one print job to another, depending on the number of objects, their sizes, and their arrangement within the build volume. It might be smaller than the build volume due to factors like print orientation and support structure requirements.

The build volume determines the maximum size of the objects that can be created with the printer. It provides an upper limit to the size of individual prints and is a critical specification when choosing a printer for specific applications. This is the most relevant consideration!

Resolution and other parameters

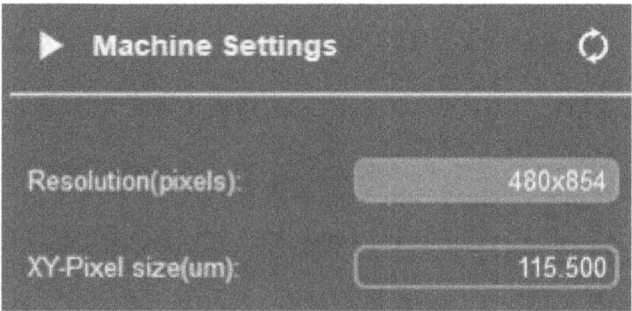

FYI, SLA printer has a resolution parameter. In those SLA printers that use a laser to cure the resin, the laser spot size is a critical determinant of resolution. The laser cures the resin layer by layer, and the smaller the laser spot, the finer the detail that can be achieved. A smaller laser spot size allows for more precise control over the curing process, enabling the printer to create intricate details and smooth surface finishes. The precision of the laser focus also affects the spot size. A well-focused laser beam will have a smaller spot size, leading to higher resolution.

In SLA printers that use Digital Light Processing (DLP) or Liquid Crystal Display (LCD) screens to project an image of each layer, the resolution is determined by the number of pixels in the projector or screen. Higher pixel density in the projector or LCD screen translates to higher resolution, as each pixel represents a smaller area of cured resin. The sharpness and clarity of the projected image directly affect the resolution of the printed layers.

Our demo printer is LCD-based. If the LCD unit becomes scratched, it will directly compromise the printer's ability to produce accurate and reliable parts because the LCD acts as a mask, controlling where UV light passes through to cure the resin. Any scratch acts as an optical defect that distorts the light path, typically resulting in either consistent, repeating defects on every print or complete print failures due to insufficient light penetration. The most common outcome is that every print will show the same defect in the exact same location on the build plate. The scratch scatters or blocks the UV light, meaning the resin in that specific area will not receive the correct dose of energy to cure properly. Consequently, you will see thin lines, gaps, or a general weakness running vertically through your model at the site of the scratch. These lines often result in partially delaminated layers, where the uncured resin fails to bond, leaving a visible split or seam on the object's surface. In more severe cases, deep scratches can cause the print to fail entirely at that point. If the light is blocked completely, no resin will cure, leading to holes that go entirely through the model or causing thin, unsupported sections to break off mid-print. On the other hand, if the scratch refracts the light in unpredictable ways, you might see unwanted, wavy deformations or small, hard bumps of cured resin in areas that should be smooth, as scattered light inadvertently cures adjacent resin.

It is important to note that the LCD screen in a resin printer is considered a consumable part. Over time, in addition to physical scratches from cleaning or removing hardened debris, the screen will naturally degrade from UV exposure, becoming cloudy or less transparent. This produces symptoms very similar to scratches, such as a general loss of detail and weak layer bonding. Because of this, many manufacturers have started including a replaceable screen protector on their printers, which is a transparent film or glass plate that sits on top of the actual LCD. If you scratch this protector, the problem is easily solved by simply peeling it off and applying a new one, saving the expensive LCD panel underneath from damage.

The XY resolution refers to the smallest feature size that can be printed in the horizontal plane (width and depth). It is defined by the smallest movement the printer can make in the X and Y directions and is directly tied to the laser spot size or the projector's pixel size. For laser-based SLA printers, XY resolution is a function of the laser spot size and how accurately the laser can be moved across the resin vat. For DLP and LCD printers, XY resolution is determined by the pixel size in the projected image.

The resolution is printer specific. For our demo, these are the resolution values:

XY Resolution: 0.1155mm, 854*480p
Z Axis Resolution: 0.01mm
Layer Resolution: 0.01~0.2mm

There are some other parameters you may want to look into.

Layer Thickness(mm):	0.050
Normal Exposure Time(s):	8.000
Off Time(s):	1.000
Bottom Exposure Time(s):	60.000
Bottom layers:	3

Layer thickness refers to the thickness of a single layer of resin that is cured in one cycle of the printing process, typically measured in micrometers (μm). Thinner layers lead to higher resolution prints with finer detail and smoother surfaces, as there are more layers per given height. Conversely, thicker layers reduce print time but may result in less detailed and rougher surfaces.

Normal exposure time is the amount of time that the resin is exposed to the curing light (laser or UV light) for each layer during the main printing process. The correct exposure time ensures that the resin solidifies adequately without overcuring, which can lead to unwanted curing beyond the desired layer thickness, or undercuring, which can cause weak layers and poor adhesion.

Adjusting exposure time involves finding a balance where the resin is sufficiently cured without excessive curing that might lead to increased print times or inaccuracies. Too long an exposure can cause excessive curing that affects neighboring areas, while too short an exposure can result in incomplete curing. The ideal normal exposure time varies depending on the resin used, its photopolymerization properties, and the printer's light source intensity. It may also need to be adjusted for different layer thicknesses.

Bottom exposure time, on the other hand, is the duration for which the resin is exposed to the curing light for the initial few layers at the beginning of the print. These layers form the base of the print and require a different exposure setting to ensure a strong foundation. The primary goal is to always ensure that these layers cure very solidly and adhere firmly to the build platform, providing a stable base for the rest of the print. Longer exposure times for the bottom layers are necessary because these layers must have a higher degree of adhesion to the build platform to prevent the print from detaching during subsequent layers' curing. It also helps to compensate for the lack of support from previous layers. Note that typically 3 to 10 layers are designated as bottom layers, depending on the print settings and object size.

Note that:
- Reducing layer thickness improves resolution but requires more layers, increasing print time.
- Adequate bottom exposure time ensures strong adhesion of the initial layers to the build platform, preventing print failures.

- *Different resins have different curing properties, requiring adjustments in exposure times to achieve the desired print quality and mechanical properties.*

FYI, the original Stereolithography technology, invented in the 1980s, uses a laser directed by a movable mirror, known as a galvanometer, to trace and cure the cross-section of each layer point by point. Because the laser draws each layer sequentially, the print speed is determined by the total surface area of the object, making it slower than other resin technologies for printing large, solid layers. However, this method offers exceptional geometric precision and produces very smooth surface finishes, as the laser can trace fine details without the pixelation effect seen on some other printers. Traditional laser-based SLA is often the go-to choice for applications where high detail and accuracy are critical, such as intricate dental models, jewelry patterns, and engineering prototypes. Formlabs has also developed an advanced version of this process called Low Force Stereolithography, or LFS, which uses a flexible tank to reduce peel forces on the part, improving reliability and surface quality further.

The other primary alternative is DLP technology. Instead of a laser, a DLP printer uses a digital projector screen to flash a single image of an entire layer onto the resin vat all at once. This means a complete layer is cured in a single exposure, making DLP significantly faster than laser-based SLA, with print speeds determined by the object's height rather than its surface area. DLP is well-suited for high-volume production runs and batch printing of smaller parts. While early DLP systems could suffer from pixelation or lower resolution at the edges of the build plate, modern systems have largely addressed these issues and are widely used in dentistry and jewelry for their balance of speed and quality.

Finally, there are also more specialized and advanced resin technologies. Carbon's Digital Light Synthesis, or DLS, is a notable evolution that uses a process involving oxygen-permeable optics to create a continuous, dead-zone where resin does not cure, allowing for much faster, truly continuous printing and enabling new types of production-grade elastomeric materials. Other innovative approaches include BCN3D's Viscous Lithography Manufacturing, or VLM, which is designed for very high-viscosity resins and can support multi-material printing, representing ongoing efforts to push beyond the capabilities of standard SLA, DLP, and MSLA systems.

Preparing the model

In the world of 3d modeling and printing, the Export function is used for converting your creation into mesh based file that can be further processed for 3d printing. The most popular mesh based file format for 3d printing is STL. In Windows 10/11 the 3D viewer feature can open and view STL files.

To print a model, first of all you must export your 3d model to STL. Then you open up the STL file from the slicer. The slicer takes the STL file and prepare a special Gcode file that can be printed.

The first section of the file contains a bunch of instructions.

```
1   ;FLAVOR:RepRap
2   ;TIME:1232
3   ;Filament used: 2.69509m
4   ;Layer height: 0.4
5   ;Generated with Cura_SteamEngine 0.0.0-master
6   M104 S200
7   M109 S200
8   ;Sliced at: Tue 01-09-2020 22:04:65
9   ;Basic settings: Layer height: 0.4 Walls: 1.2 Fill: {fill_density}
10  ;Print time: 00:20:33
11  ;Filament used: [2.7]m [8.038241874999999]g
12  ;Filament cost: [0]
13  ;M190 S60 ;Uncomment to add your own bed temperature line
14  ;M109 S200 ;Uncomment to add your own temperature line
15  G21         ;metric values
16  G90         ;absolute positioning
17  M82         ;set extruder to absolute mode
18  M107        ;start with the fan off
19  G28 X0 Y0   ;move X/Y to min endstops
20  G28 Z0      ;move Z to min endstops
21  G1 Z15.0 F3000 ;move the platform down 15mm
22  G92 E0                  ;zero the extruded length
23  G1 F3000 E3             ;extrude 3mm of feed stock
24  G92 E0                  ;zero the extruded length again
25  G1 F3000
26  ;Put printing message on LCD screen
27  M117 Printing...
28  ;LAYER_COUNT:48
29  ;LAYER:0
30  M107
31  G0 F3600 X16.687 Y-5.049 Z0.3
```

The actual data section is like a map full of coordinates.

P. 48

```
1287    G1 X12.45 Y21.658 E246.95822
1288    G1 X12.45 Y21.229 E246.96584
1289    G1 X12.45 Y20.89 E246.98344
1290    G1 X12.45 Y20.878 E246.98424
1291    G1 X12.45 Y-20.866 E249.76106
1292    G1 X12.45 Y-20.997 E249.76977
1293    G1 X12.45 Y-21.32 E249.79126
1294    G1 X12.449 Y-21.658 E249.81374
1295    G1 X12.634 Y-21.321
1296    G1 X12.796 Y-20.997 E249.82912
1297    G1 X12.847 Y-20.877 E249.83722
1298    G1 X12.972 Y-20.585 E249.85835
1299    G1 X13.056 Y-20.352 E249.87482
1300    G1 X13.136 Y-20.096 E249.89267
1301    G1 X13.203 Y-19.841 E249.9102
1302    G1 X13.263 Y-19.552 E249.92984
1303    G1 X13.304 Y-19.256 E249.94972
1304    G0 F7200 X13.321 Y-19.057
1305    G0 X12.996 Y-19.784
```

The Gcode file is in fact a text file which can be viewed with Notepad and the like. The file can be saved to a SD card directly through Cura (or you can manually copy the file to the SD card). Almost every 3D printer has a SD card slot for reading Gcode files. Some even allow you to connect directly to the computer via USB.

Our example unit supports the use of USB flash drive for printing. HOWEVER, it uses a proprietary pw0 format instead of the standard gcode file. Not all slicer software support this format.

For SLA printing, using an SD card or USB flash drive is strongly recommended over a direct USB connection to a computer. While both methods can technically work, printing directly from a PC introduces several risks that can compromise your prints, while offline storage offers greater reliability and safety.

The most significant advantage of using an SD card is the independence it provides. When you print from a memory card, the printer reads the sliced file and executes it entirely on its own, meaning the print job is not reliant on your computer staying awake, avoiding updates, or experiencing any other interruptions. Many users report that the SD card method fails the least because there are fewer variables, such as Wi-Fi cutting out, network issues, or the computer updating and restarting in the middle of a long print that may last forty hours or more. Many printers also include an SD card slot as standard functionality, and some models are explicitly marketed as having this feature.

Direct USB printing, conversely, involves your computer sending commands to the printer line by line in real time. If your computer crashes, goes to sleep, or even suffers a momentary CPU spike that delays command transmission, the print can pause or fail. For resin printing, where timing is critical because each layer requires a precise exposure duration, such delays can ruin a job. Users have described experiencing this firsthand, noting that if the PC crashes or reboots, the printer will stop moving. Beyond print quality issues, there can also be safety concerns, as the heating elements might continue running even if the print process stalls.

That said, there are advanced exceptions. For users who want the convenience of sending files wirelessly without using a memory card, a dedicated device like a Raspberry Pi running OctoPrint can be a good solution. This setup still uses a USB cable, but because the small, low-power computer is dedicated solely to managing the printer, it is far more stable than using a general-purpose PC. Similarly, more sophisticated firmware like Klipper offloads processing to a host computer by design, making USB printing viable in those specific configurations. For the standard SLA user with most common printers, however, the SD card method remains the most reliable and trouble-free approach, as it completely removes your computer from the equation and allows your printer to work safely and independently without the risk of a digital glitch ruining a multi-hour print.

The brain of the printer

G-code is a special computer language. It tells the 3d printer how to print things. In particular it has commands carrying a bunch of assigned movement and action the 3d printer would follow.

As mentioned earlier, a 3d printer creates a 3D object by adding material on a layer-by-layer basis. Slicing is all about cutting a 3d object into horizontal 2d layers so that the object can be printed one slice at a time. It is like piling up a bunch of paper sheets each with different shapes so to form a 3d object.

The Gcode file has information on all the slices. The "brain" inside the 3D printer can interpret the information accordingly. Do note that some slicers also support the X3G format, which is less popular than Gcode.

There is a set of microcontroller and additional circuit boards /devices inside every printer. These electronic components together read the Gcode file and instruct the print head to print accordingly.

Aka MCU, a microcontroller is like a low cost simplified version of your regular computer, or some sort of system on a chip. Those found on a 3D printer are optimized for controlling the various parts so to 3d print your models.

The printer controller circuitry needs to offer high current capabilities as well as support for MOSFETs, movement systems, LCD screen and printer firmware ...etc.

Older systems are 8 bit based. 32 bit boards are now very affordable! Some can provide power to the components directly, while some require help from special driver circuitry. If for whatever reason the circuitry is damaged, you should always replace the bad one with one of an exact same model. Arduino seems to be quite popular these days. Different Arduino boards may be equipped with different chips, even though the Atmel (now owned by Microchip) controllers are the most popular ones. ATmega8, ATmega168, ATmega328, ATmega1280, and ATmega2560...etc. The primary difference that really has a meaning to a beginner is the amount of memory onboard. The Uno is the most popular one for starters. The Arduino Mega is like a more powerful version of the Uno.

The Arduino board alone does not offer 3d printing logic as it cannot talk to other components directly. There has to be a middleman in between.

The compartment that houses the board is small so the power supply to the printer has to stay outside as a separate unit. There is usually a dedicated connector for it.

Infill

The shell is the outer wall of the object. Infill, on the other hand, is the partially hollow interior. A "thicker" wall is of course better structural-wise. The infill is invisible to you. Typically the infill has a pattern of grid or honeycomb. Fill Density is a parameter in percentage you can specify when doing FDM printing. Higher density means more rigid structure. It will also consume more printing material. A density of 0% means hollow. A 20% density is typically good for printing a prototype.

But why no Infill settings for SLA printers?

A resin printer develops a model which is submerged in a vat of resin. If a slicer was to enable traditional Filament style infill within a resin model, all you would get would be a model with trapped, uncured resin in all of the infill spaces which could leak uncured resin.

SLA printed models are always 100% infill density-wise by default.

Some slicers may offer the option to hollow out a model and add holes to allow air into the void during printing so to allow the uncured resin to drain out of the void as the model prints. The slicer that comes with our demo unit seems not.

Proper support

Supports are sometimes needed in SLA printing since they provide structural stability during the printing process and help in accurately producing overhangs, bridges, and intricate geometries that would otherwise be impossible to print.

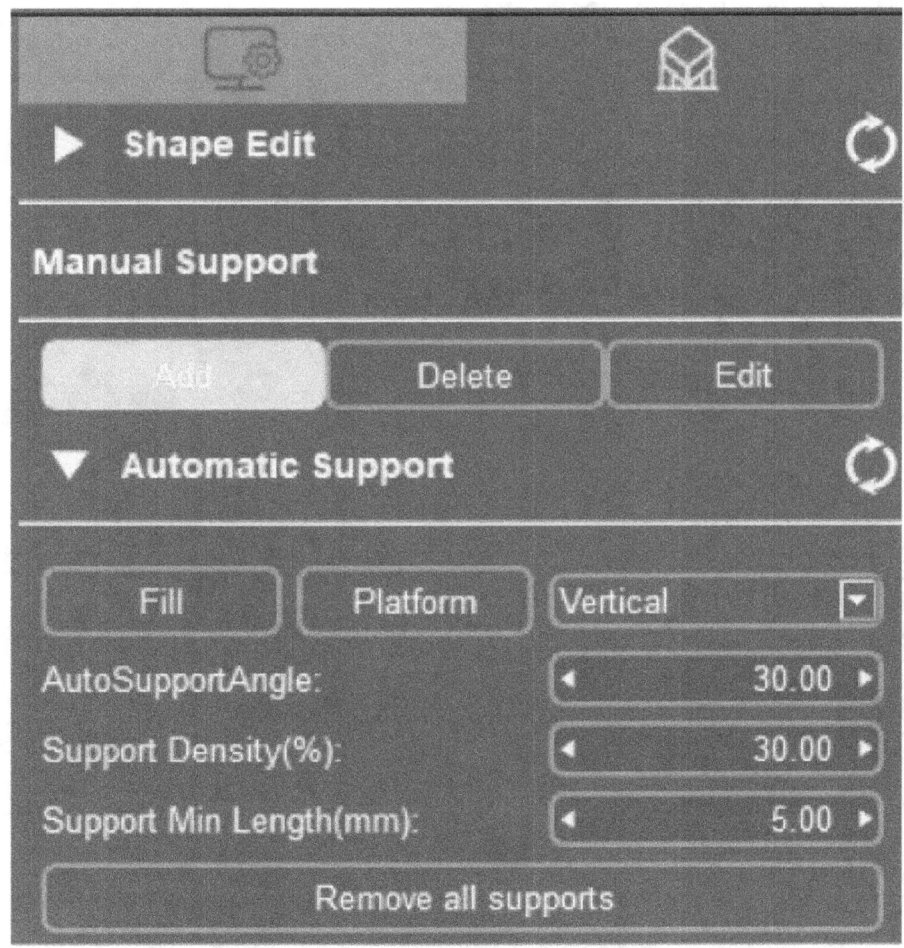

Our slicer can generate automatic support if needed. The default settings are fine for beginners. I would not remove the supports ...

supports are critical when you want to print something larger....
Keep in mind, supports are necessary for overhanging parts that extend beyond a certain angle from the vertical. Without supports, these features would sag or fail during printing. Simply put, they act as a scaffold, holding the overhanging parts in place until they are fully cured. Also, during the printing process thermal stresses can cause warping or distortion, especially in large or thin structures. Supports help in distributing these stresses, ensuring that the printed part maintains its intended shape and dimensions.

Warping...

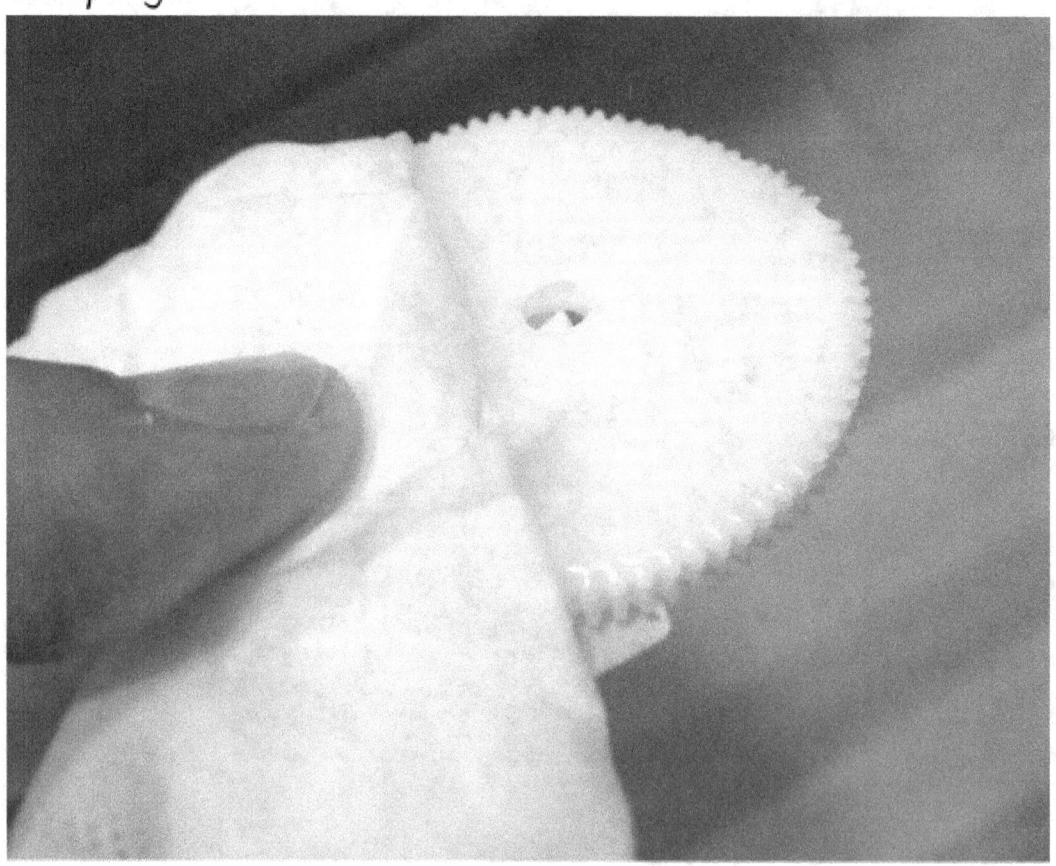

And, supports help in stabilizing the part during the printing process, reducing the risk of shifting or falling. They provide additional anchor points, preventing the part from moving or being displaced by the resin's recoating process.

You need to know the concept of YHT. Both letter Y and letter T have overhangs at the top. The letter H, in contrast, has a bridge in the middle. Generally, overhangs less than 45 degrees may not require support. Bridge less than 5mm in length may also not require support.

Most modern slicer software are smart enough to prepare the necessary support for you. If not, you may need to include some support elements in your model design. In fact this is a recommended approach if your aged slicer software does not really have the necessary intelligence to generate proper support for complicated models.

The specific types of supports an SLA printer can produce fall into a few main categories based on their shape and structure. The most common type is the pole or pillar support, which resembles a series of thin columns or rods. This shape is widely used because it is material-efficient, contacting the model only at the very tip and preserving the surface quality of the final print. Advanced slicers can also generate organic or tree-like supports, which use branching structures that grow from the sides of the model. These not only use less material but are also easier to remove and leave behind fewer noticeable scars on the model's surface. Another variation is the mesh or grid support, which is composed of intersecting thin sheets that form a dense lattice. This type fully encapsulates the bottom of a model, providing maximum stability for large, flat surfaces, and its thin walls make it relatively easy to detach.

Beyond these visual styles, the software that prepares your print for an SLA machine offers several intelligent support styles. Grid supports are very stable and have a well-defined structure, making them reliable for general use. Snug supports are designed to conform perfectly to the shape of the specific overhang they are supporting, reducing the amount of contact with the model's main walls. In industrial-grade software, you may also encounter sophisticated options like bar supports, polyline supports, or even full volume supports, which fill internal cavities to brace the model from the inside out.

Regardless of the style, all SLA supports are temporary. After the print is finished, they must be removed in a process called post-processing. Because they are integral to the print, removing them leaves small marks or pockmarks on the surface where they were attached. However, because these supports are specifically designed for resin printing, they are optimized to balance holding strength with ease of removal, making the post-processing step a manageable part of the SLA workflow.

Filling, Printing and cleaning

Adding resin to your SLA printer properly means filling the resin tank or vat. Start by cleaning it if it contains old resin or debris. You can use a plastic scraper to remove cured resin bits and a paper towel with isopropyl alcohol to clean the tank/vat thoroughly. Shake the resin bottle for a couple of minutes to ensure the resin is well-mixed. Open the bottle carefully and pour the resin slowly into the resin tank to avoid forming bubbles. Fill the tank to the recommended level, which is usually indicated by a maximum fill line or indicator in the printer. After pouring the resin, check for bubbles.

If bubbles are present, let the tank sit for a few minutes or gently stir the resin with a clean plastic spatula to release them.

Follow the printer's instructions to level the build platform (this will be discussed further in the next chapter). Insert the USB storage or SD card or whatever into the printer then choose Print from the LCD screen.

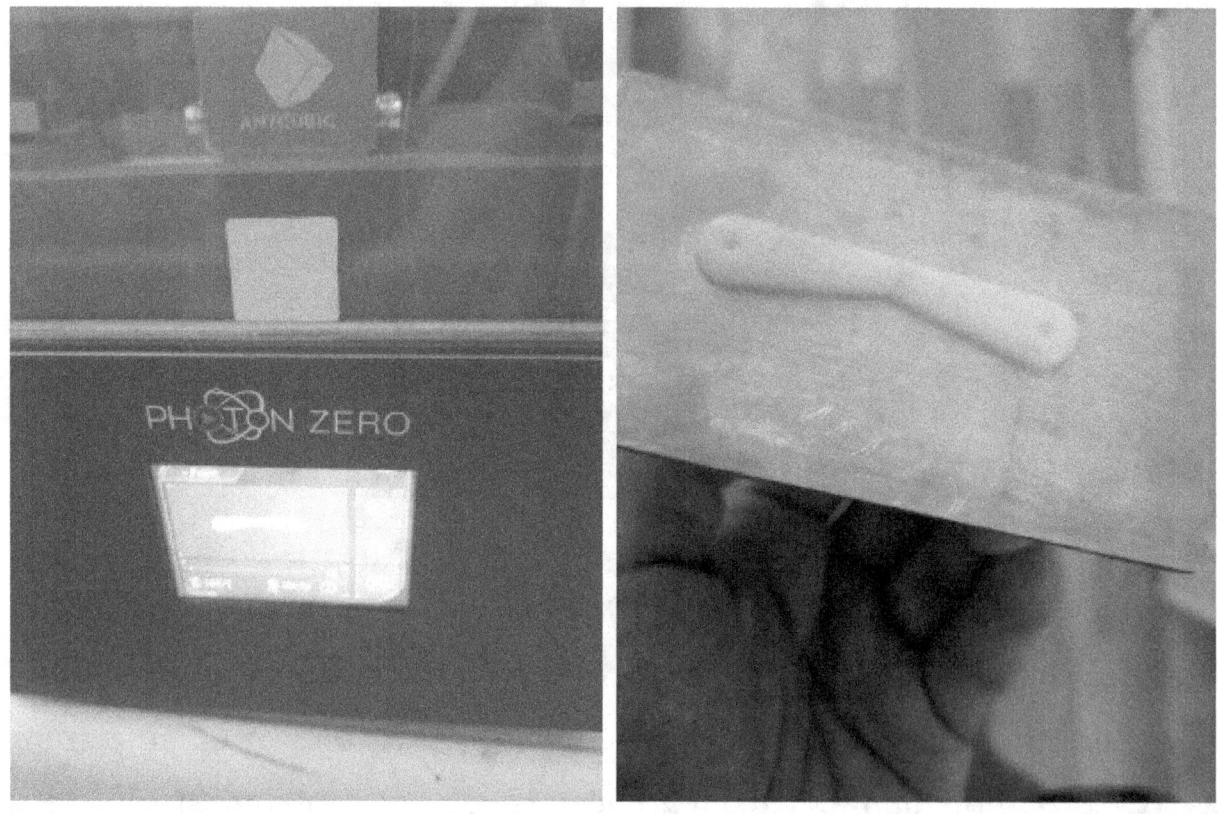

Once the print is complete, remove the build platform and carefully detach the printed object using a scraper.

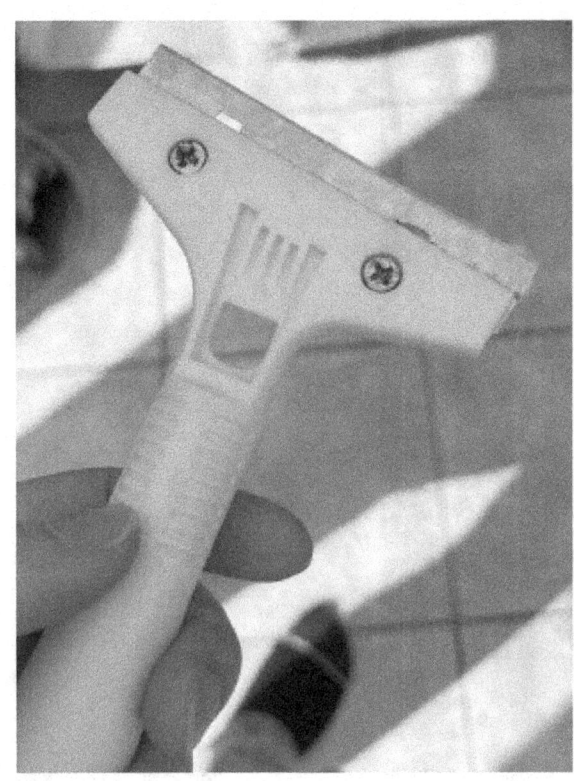 

Rinse the print in isopropyl alcohol to remove any uncured resin. Some printers come with a wash station specifically for this purpose. If your resin type requires it, cure the print in a UV curing station to fully harden it.

If you plan to switch resins or won't be printing for a while, clean the resin tank thoroughly. Filter the resin back into the bottle using a resin filter or funnel to remove any debris. Store the resin bottle in a cool, dark place with the cap tightly sealed to prevent exposure to light and air.

On our demo printer the vat can be easily removed for cleaning.

 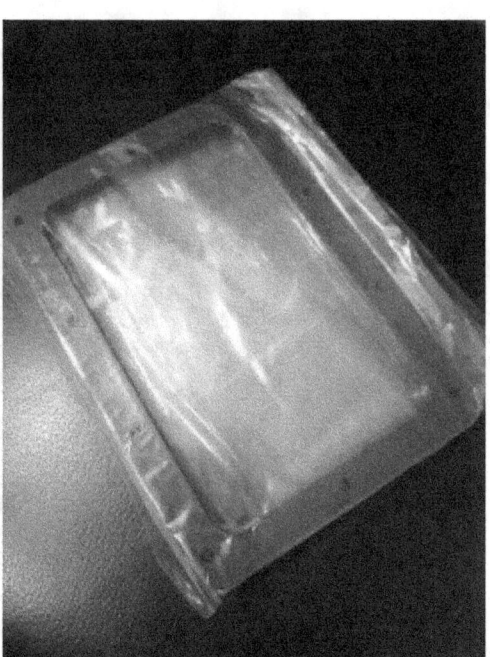

The bottom of the vat tray can actually be replaced. You may want to replace it if it has scratches on it (it should remain clear so the laser light can reach the resin evenly and properly).

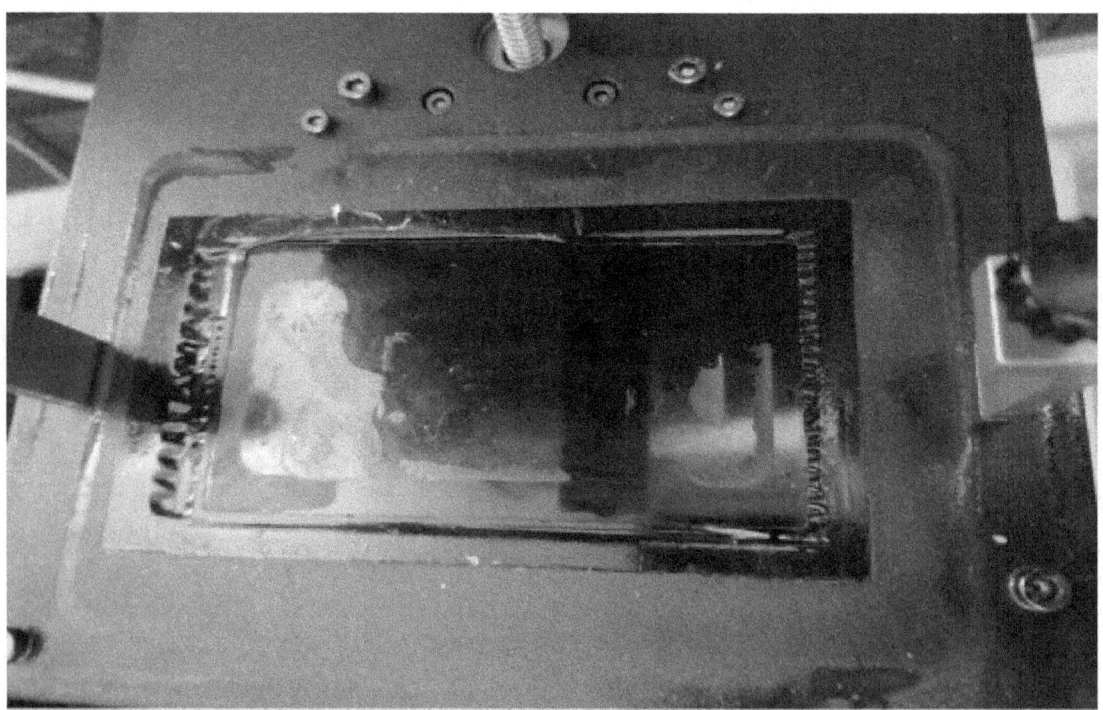

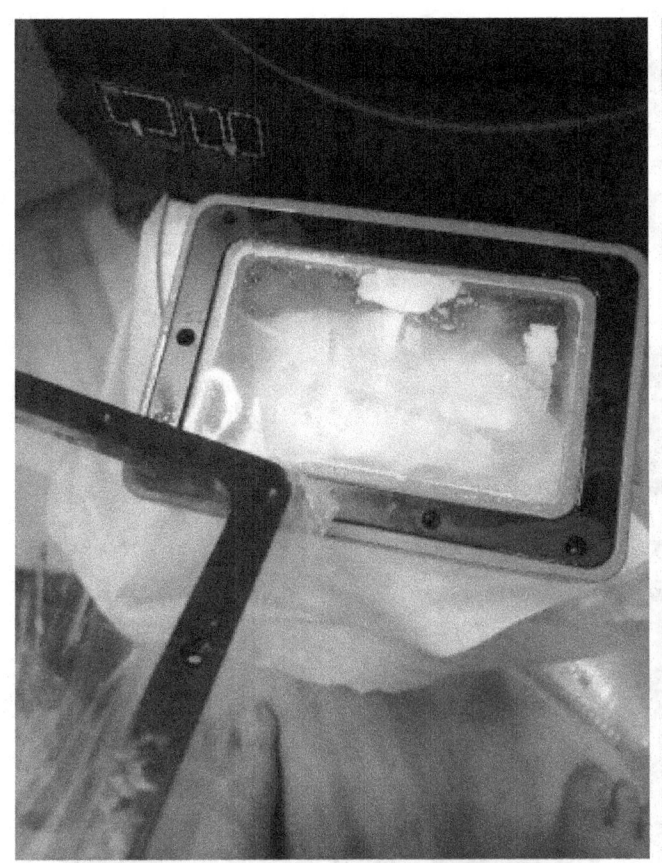
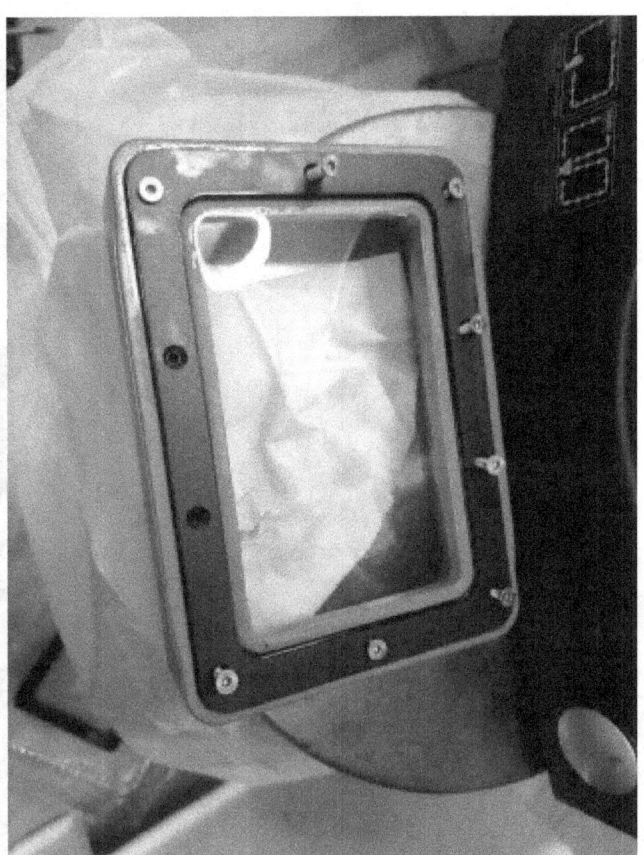
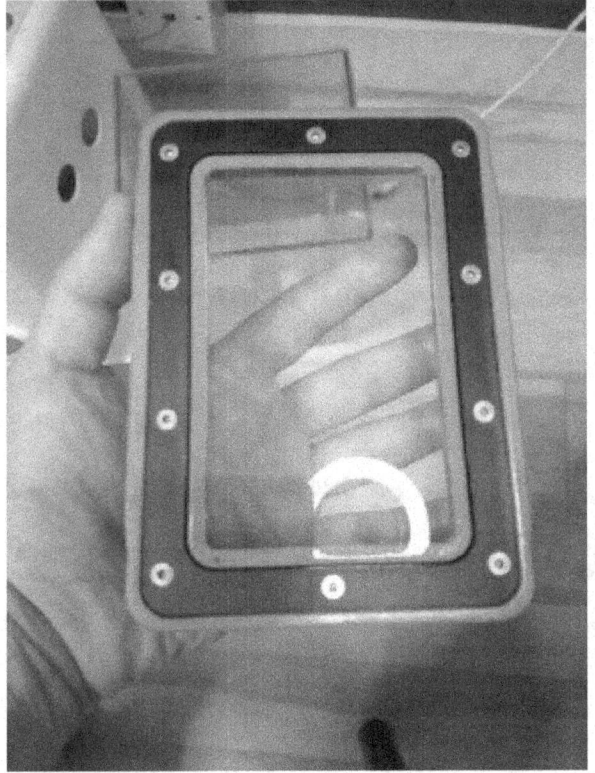
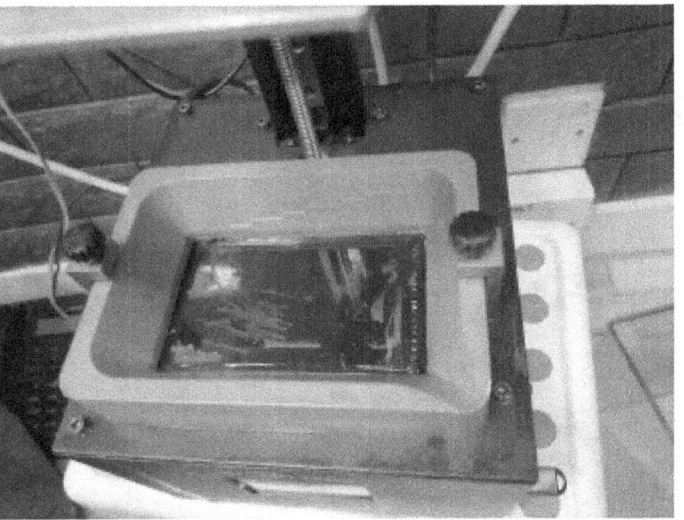

In fact, if you change to another type of resin you may want to replace the vat too. It will also be necessary to clean the LCD underneath the vat. But be careful not to damage it.... scratched screen here can lead to all sorts of terrible print outcomes...

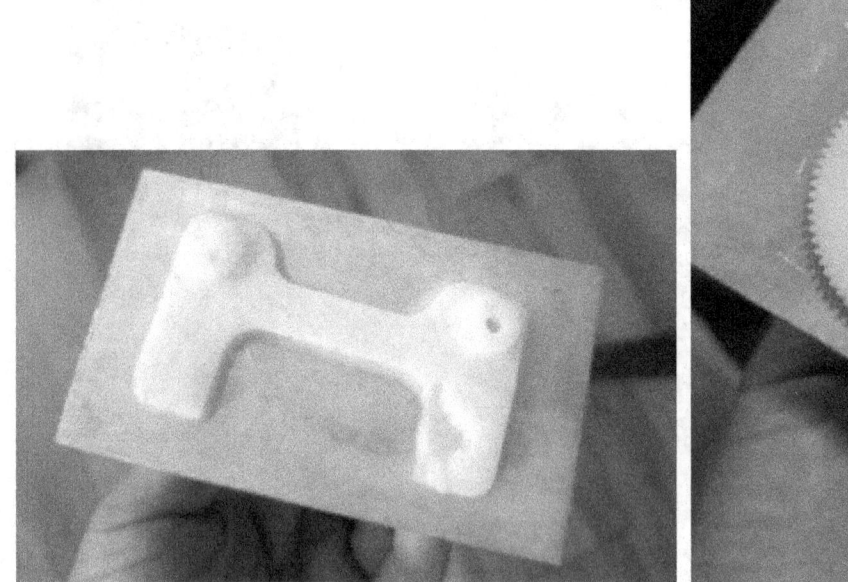

The demo unit we have here has a replaceable LCD module too. In fact, many name branded printers are built on modules where replacement is easy as long as parts are available.

Resin crystallization could be a big problem. Resin crystallization occurs when the liquid photopolymer resin begins to form solid crystals, disrupting the printing process. This problem can arise due to several factors, such as improper storage conditions, contamination, or extended exposure to light or heat.

Improper storage is a common cause. Resins must be stored at specific temperatures, typically between 15°C to 25°C (59°F to 77°F). Extreme temperatures can lead to crystallization. Additionally, since resin is photosensitive, exposure to light (whether natural or artificial) can initiate unwanted curing and crystallization. Therefore, resins should be kept in opaque, tightly sealed containers to prevent light exposure.

Contamination is another factor. Dust, dirt, or other foreign particles can act as nucleation points for crystal formation. Mixing different types or even batches of resin can sometimes cause chemical reactions leading to crystallization as well. And, extended exposure to fluctuating temperatures or light can also cause problems. Resin left in the printer's tank for prolonged periods, especially if the printer is not in a controlled environment, is prone to crystallization. Similarly, resin in partially used bottles can start to crystallize over time if not sealed properly.

Detecting resin crystallization involves looking for visible crystals in the resin tank or at the bottom, appearing as cloudy, grainy, or crystalline structures within the liquid resin. Print quality issues are a telltale sign, with problems such as incomplete curing, rough surfaces, and weak structural integrity of printed objects. Additionally, crystallized resin may become thicker and more viscous, making it difficult to pour or filter, and can clog the printer's resin tank or dispensing mechanism.

To address and prevent resin crystallization, proper storage is essential. Always store resin in a cool, dark place, within the manufacturer's recommended temperature range, using opaque containers to protect it from light. Regular maintenance of the printer, including cleaning the resin tank and filtering the resin to remove any formed crystals, is also crucial. Ensuring the workspace and tools are clean and avoiding mixing different resins or batches can help prevent contamination. Using fresh resin and avoiding leaving it unused for long periods can also reduce the risk of crystallization. If crystallization is detected, filtering and cleaning the resin before printing again is necessary to maintain the quality and performance of the SLA printer.

One interesting question: can you print without the lid / cover? In fact your printer may have a detect function which detects whether the lid / cover is on (our demo unit has such function too).

Technically, you can leave the cover / lid off. HOWEVER, the resin used in SLA printing cures and hardens when exposed to specific wavelengths of UV light. Without the lid or cover, ambient light, especially UV light from sunlight or artificial sources, can prematurely cure the resin in the tank. This premature curing can cause clumping and clogging in the resin tank, leading to print defects or complete failure of the print job.

Additionally, without the lid, the resin tank is exposed to dust, debris, and other contaminants. These contaminants can settle into the resin, causing surface defects on the printed object and poor layer adhesion, which compromises the structural integrity of the final print.

There are also safety concerns when printing without a lid. SLA resin emits fumes that can be hazardous if inhaled over long periods. The cover helps contain these fumes and typically works in conjunction with ventilation systems designed to minimize exposure. Furthermore, exposure to uncured resin can cause skin and eye irritation or allergic reactions. An open setup increases the risk of accidental spills or splashes, making it more hazardous to handle the resin.

Temperature control is another critical aspect of the printing process. The resin's viscosity and curing properties are affected by temperature changes. An open setup might not maintain a stable temperature, leading to inconsistent curing and print quality issues. If the resin becomes too cool, it can become more viscous, making it harder for the printer to dispense and cure properly.

The lid or cover also helps to maintain a stable printing environment. A closed environment minimizes the effects of air drafts, temperature fluctuations, and humidity changes, all of which can impact print quality. Moreover, covers can help dampen vibrations or movements that might affect the printer's precision.

Delamination and other defects

Delamination refers to the failure of the printed layers to bond properly, leading to separation or peeling between the layers of the object. This issue typically arises during the printing process when the cured resin from one layer fails to adhere well to the resin of the layer below it. As a result, the object may experience weak points or fractures where the layers do not fuse adequately, compromising the structural integrity of the final print.

Delamination can occur for various reasons, including improper curing settings, insufficient exposure times, or issues with the resin itself. If the resin is not cured for the correct amount of time during the printing process, the layers may not bond fully, leading to weak adhesion. In addition, if the printing environment (such as the temperature or humidity) is not optimal, it can affect how the resin reacts and cures, further contributing to the problem.

The problem is more likely to occur in prints with complex geometries or thin, fragile structures, as these can put additional stress on the layers during printing. In some cases, delamination may also result from the printer's mechanical issues, such as inaccuracies in layer alignment or excessive movement during the printing process. Extended pauses during the print process can lead to delamination after the print resumes. Damage or scratches on the resin tank may also block the laser at that location, thus preventing the resin directly above it from curing properly.

Preventing delamination typically involves fine-tuning the exposure settings, using a suitable resin, and ensuring proper printer calibration.

If the LCD screen of your SLA printer is not clear or has scratches, it can lead to several issues that affect the print quality and overall performance of the printer. Note that the LCD is physically stressed during every print. The process of the build plate lifting and the FEP film peeling away creates small but repeated forces on the screen, as the FEP is pressed directly against it. The screen also generates heat from both the UV LED array beneath it and the absorption of UV light itself. This combination of thermal cycling and mechanical stress contributes to its gradual decline.

Manufacturers typically rate LCD screens for a specific number of working hours, often between 400 and 2,000 hours depending on the quality of the screen and the printer. A heavy user printing continuously might reach this lifespan in just a few months, while a casual hobbyist might get a year or more. Signs that your LCD is nearing the end of its life include prints failing consistently in certain areas of the build plate, vertical lines or bands appearing on every print, a noticeable drop in detail and sharpness, or the need to significantly increase exposure times to achieve proper curing.

The good news is that most resin printers are designed with this in mind. LCD screens are typically modular and user-replaceable, and manufacturers sell replacement screens directly. Many modern printers also feature a sacrificial glass or plastic screen protector that sits on top of the actual LCD. This protector absorbs minor scratches and resin spills, and replacing it is much cheaper and easier than replacing the LCD itself.

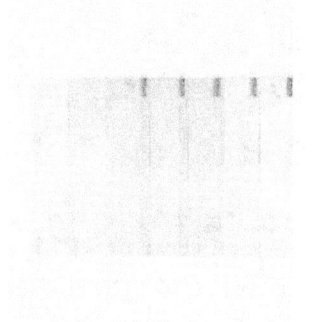

ALadrs Screen Protector for LCD Resin 3D Printers

£12.99

a Amazon.co.uk - ...
Free delivery

KOYOFEI 5PCS Screen Protector for LCD Resin 3...

£14.99

a Amazon.co.uk - ...
Free delivery

ALadrs Screen Protector for LCD Resin 3D Printer

£11.99

a Amazon... & more
Free delivery

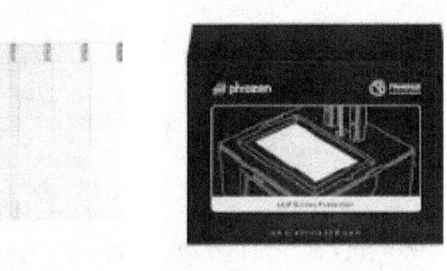

Phrozen LCD Screen Protector

£8.44

phrozen3d.com

LCD screen contamination

LCD screen contamination in SLA printers is primarily caused by resin spills, improper handling, and environmental factors. Removing or replacing the vat carelessly can lead to resin spilling onto the screen. Over time, wear and tear or accidental damage can cause the vat to leak resin onto the screen. Filling the resin vat beyond its recommended capacity increases the risk of spills during operation.

To clean a contaminated LCD screen of an SLA printer, it's important to proceed carefully to avoid causing damage while effectively removing resin or other contaminants. Start by turning off the printer and unplugging it to ensure safety and prevent accidental activation. Wear nitrile gloves to protect your hands and prevent further contamination during the cleaning process.

Remove the resin vat to expose the LCD screen. If the vat contains uncured resin, carefully drain it into an appropriate container for reuse or disposal. Once the screen is exposed, use a dry microfiber cloth to gently blot any excess liquid resin from the surface without spreading it further.

Dampen a microfiber cloth or a cotton swab with a small amount of isopropyl alcohol (IPA) with a concentration of 90% or higher. Avoid pouring IPA directly onto the screen, as excessive liquid can seep into the printer and cause damage. Use the damp cloth or swab to wipe the screen gently in a circular motion, focusing on the contaminated areas. For corners and edges, a cotton swab can help clean spots that are hard to reach.

If cured resin is present on the screen, a plastic scraper can be used to carefully lift it off. Do not apply excessive pressure, as this could scratch or damage the screen. For stubborn spots, place a cloth dampened with IPA over the area and allow it to sit for a few minutes to loosen the resin before attempting to remove it.

After cleaning, use a clean, dry microfiber cloth to remove any streaks or remaining residue. Inspect the screen for any remaining contaminants, and repeat the process if necessary. Once the screen is clean, allow it to air dry completely before reassembling the printer and powering it on.

Be cautious not to use harsh chemicals like acetone, which can damage the screen coating, and avoid abrasive materials that could scratch the surface. Regular inspection and maintenance of the LCD screen can help prevent significant contamination and ensure consistent print quality over time.

Retraction and exposure

In SLA printing, Z retract speed refers to how quickly the head moves vertically when pulling the resin from the build surface or layer. A longer retract speed can pull too much resin from the vat, leading to excess resin being displaced from the build area. This may result in uneven curing or problems with layer bonding, affecting print quality.

Longer retraction times add unnecessary delays in the printing process, leading to longer print times overall, without improving print quality. In some cases, an overly long retract speed could cause the resin to drip from the print head or nozzle, leading to blobbing or inconsistent layer formation.

```
Z Lift Distance(mm):      ◄      6.00  ►
Z Lift Speed(mm/s):       ◄      3.00  ►
Z Retract Speed(mm/s):    ◄      3.00  ►
```

FYI, retraction is related to the lifting distance and lifting speed. After each layer is cured, the build plate lifts to separate the print from the bottom of the resin vat, and this process is critical for success. Lifting distance is how high the build plate moves up after each layer, with a standard distance typically falling between 6 and

8 millimeters. If this distance is too short, the print may not fully peel away from the FEP film, causing layer shifts or print failures. Lifting speed is how fast the plate moves up, and a slower speed, such as 60 to 90 millimeters per minute, reduces the peel force on the print, making it less likely to detach from the build plate or for delicate supports to break. Some printers use a two-stage lift, for example moving first at 85 millimeters per minute and then at 240 millimeters per minute, to balance gentle separation with overall print speed.

Exposure time refers to the duration for which the resin is exposed to a light source to cure and solidify each layer. Proper exposure times are critical for achieving accurate prints with good mechanical properties. Bottom layers (first few layers) often have longer exposure times to ensure strong adhesion to the build plate. ubsequent layers have shorter exposure times to balance speed and resolution. The Bottom Layer Exposure Time is typically 2-10 times longer than normal layer exposure.

▼ **Slice Settings**

Layer Thickness(mm):	0.050
Normal Exposure Time(s):	8.000
Off Time(s):	1.000
Bottom Exposure Time(s):	60.000
Bottom layers:	3

FYI, getting the exposure time right is a balance between strength and accuracy. Normal layer exposure, which is the curing time for the main body of the print, typically ranges from 1.5 seconds to 8 seconds for a 50 micron layer height, depending on the printer's power and resin properties. If the time is too low, layers will not fully cure, leading to delamination or a weak, mushy print. If it is too high, you will lose fine details, and the print may become brittle or have a rough, over-cured surface. The bottom layers, usually the first four to eight layers, require significantly longer exposure, often 30 to 60 seconds or more, to create a strong bond with the build plate and prevent the print from falling over mid-print.

Levelling

Whatever you have done to the printer, leveling has to be done after parts replacement or maintenance work, otherwise the print result will be completely off.

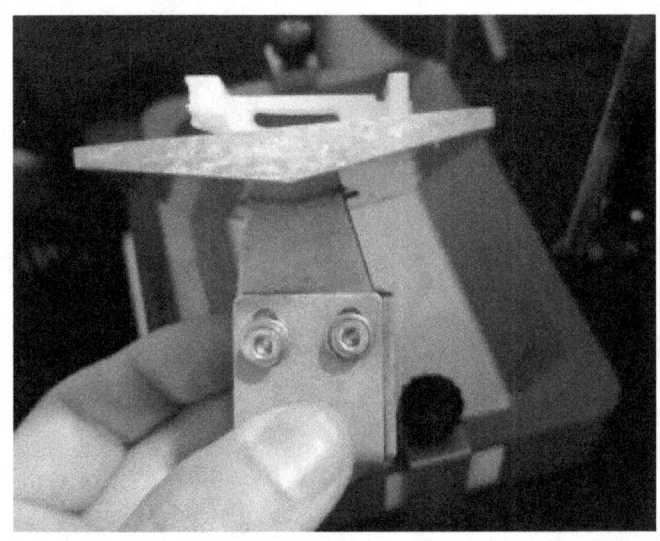

Even though you can visually perform leveling, there are more precise ways to do the job.

One good way to level the build platform of an SLA printer is by using an A4 paper. You start by turning off the printer and disconnecting it from the power source. Ensure the build platform is clean and free of any resin or debris. For easier access and visibility, remove the resin vat/tank if the design of your printer allows it (it will be difficult to do the leveling if the vat gets in the way).

Next, turn on the printer and use the controls to move the build platform to the home position, which is typically done through the printer's menu system. Take a clean sheet of A4 paper and fold it once to increase its thickness slightly. Place the folded A4 paper on the printer's base, where the resin tank usually sits.

Using the printer's controls, carefully lower the build platform until it is close to the paper, moving it incrementally to avoid any sudden movements that might damage the platform or the printer. Locate the screws or knobs that allow you to adjust the platform, which are usually situated at the corners of the build platform.

Slide the paper back and forth under the build platform and adjust the screws until you feel a slight, uniform resistance when moving the paper. Move the build platform to different areas, such as the corners and center, and check the resistance at each point, making small adjustments as necessary to ensure the platform is level across its entire surface.

Once the platform is evenly leveled, carefully remove the A4 paper from the printer and reinstall the resin tank if you had removed it earlier, making sure it is properly seated. Double-check the level by performing a test print or running the printer's built-in leveling check if it has one.

Your printer model may also have specific ways to do leveling. Take our demo unit as an example, the manual has a section on leveling (which also uses A4 paper):

5. Place a piece of A4 paper about 0.1mm thick on the curing screen. Then click on the touch screen. Wait for Z axis to descent and then it will stop automatically.

6. Finger press on top of the platform gently, and then tighten the four screws to secure the platform.

END OF BOOK

Please email your questions and comments to *editor@hobbypress.net*

Print your own parts for repair and upgrade! UpgradePARTS.com has an extensive collection of free 3d models for airsoft guns, hobby models, drones and RC cars.

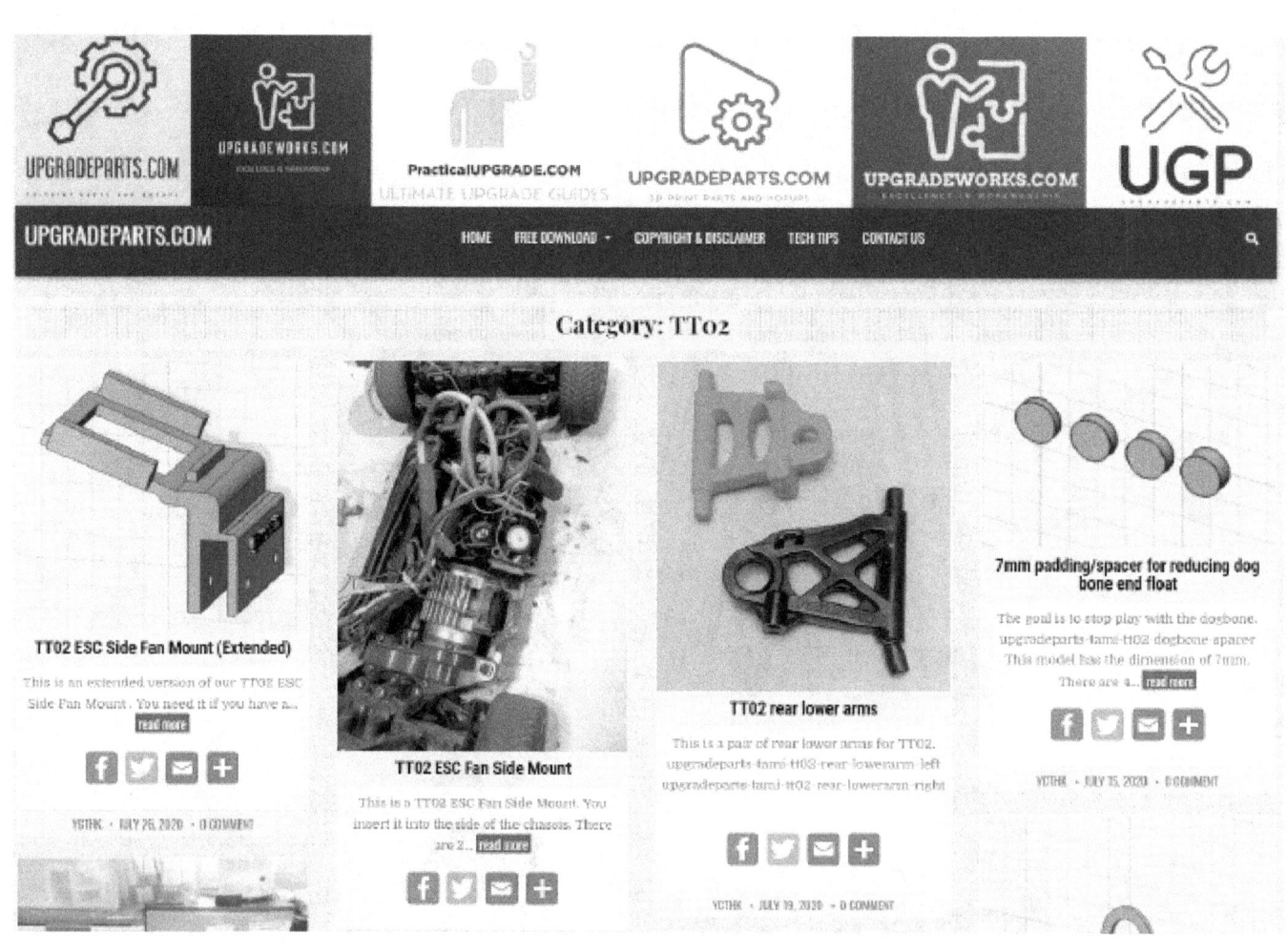

http://upgradeparts.com

www.ingramcontent.com/pod-product-compliance
Lightning Source LLC
Chambersburg PA
CBHW082238220526
45479CB00005B/1274